AUDEL™

T0048430

Audel Electrical Trades
Pocket Manual

⦚AUDEL™

Audel Electrical Trades Pocket Manual

L.W. Brittian

WILEY

John Wiley & Sons, Inc.

Library of Congress Cataloging-in-Publication Data:

Brittian, L. W.
   Audel electrical trades pocket manual / L. W. Brittian.
      pages cm
   Includes index.
   ISBN 978-1-118-08664-3 (pbk); 978-1-118-27775-1 (ebk);
   978-1-118-27776-8 (ebk); 978-1-118-27778-2 (ebk); 978-1-118-27780-5 (ebk);
   978-1-118-27789-8 (ebk); 978-1-118-27790-4 (ebk);
1. Electric industries—Vocational guidance—Handbooks, manuals, etc.
2. Electrical engineering—Vocational guidance—Handbooks, manuals, etc.   I. Title.
   TK159.B75 2012
   621.3--dc23

                                                                    2011045251

Printed in the United States of America
SKY10068658_030324

CONTENTS

INTRODUCTION

What should a technical author do for his reader? He should acquire, distill, condense, simplify, illustrate, and format information. The information in this book provides "aspirins" for the reader's electrical headaches. Hands-on personnel have little time to catch up on their work-related reading. This small book is brief, simple, and graphic. Engineering terms and technical words, when possible, have been replaced with high-school English. It presents only the "need to know" information, not the nice to know, and seldom-used book "filler"-type information.

An economy of both words and topics is exercised to help the reader get to the meat of the idea in a minimum amount of time. Authors many times attempt to cram an entire electrical library in a book that will fit in the pocket of a pair of jeans.

This is just one in a growing list of the Audel series of famous small books that are quick and easy to read. They hit the topics like a pneumatic nail gun, head on, and quick.

Audel has been publishing electrical books since 1911 and they know how to "get ''er done." This book is for those hands-on personnel who work in commercial and industrial facilities as electricians, or multicrafts maintenance personnel.

It covers the sources of many everyday electrical questions, as simply as possible.

Aware that a picture is worth a "thousand words," over two hundred illustrations have been provided to help the reader grasp the ideas quickly and easily.

Sometimes a small chart, table, or drawing can provide the needed information faster than 20 pages full of words. The reference section (Chapter 19) is packed with commonly needed reference materials.

Providing answers to basic electrical questions, this book will prove to be of value to the reader for many years to come.

L. W. Brittian

DISCLAIMER

The reader is expressly encouraged to adopt all safety precautions when working on, or near electrical equipment, specifically the Standard for Electrical Safety in the Workplace, and all company policies and procedures.

Neither the publisher nor the author make or express any warranties of any kind, including fitness for particular purpose or merchantability. The publisher and the author shall not be libel for any such special, consequential, or exemplary damages resulting, in whole or part, from the reader's use or reliance upon this material.

1

THE LAWS THAT CONTROL AND EXPLAIN ELECTRICITY

This chapter will cover the dominate type of electrical power used in commercial and industrial facilities, alternating current (AC).

Electricity is the movement of electrons too small to be seen by the human eye; it cannot be heard with our ears, or smelled by the nose. Yet if it is touched, it can result in injury or death. The most basic and important facts about electricity have been covered. Now let us consider the rest of the electrical story.

A pressure in an electrical system called voltage causes the unseen electrons to move and is measured in units called volts. As the AC voltage and current is always changing in value, it was determined that if the peak AC voltage was multiplied by .707 it would be about the same as a DC (direct current) voltage. While a meter displays 120 volts AC, the peak voltage (PV) is about 169 volts AC.

$$PV \times .707 = RMS \quad 169 \times .707 = 120 \text{ volts}$$
$$RMS \times 1.414 = PV$$

The root mean square (RMS), and not the average, is used with a continuous time waveform such as an alternating current. Since AC voltage changes polarity either 50 or 60 times (called the *frequency*, or Hertz) each second, it would be impossible to read the meter's fast-moving display with it changing 50 or 60 times each second.

Buffers built into meters slow the speed at which the meter's display changes allowing it to be read. Better quality meters display RMS values.

The quantity of those unseen electrons moving in an electrical circuit are measured in a unit called *amperes*, or amps, abbreviated with the letters A and I for current.

One ampere contains 6,250,000,000,000,000,000 electrons (or 6.242 times 10 to the 18th power). One volt is defined as the electromotive force (EMF) (pressure) that will move one ampere (quantity) of electrons through a resistance (opposition or friction measured in units of ohms) of one ohm. That is, one volt will move one amp of electrons through a resistance of one ohm.

The three main variables of interest are voltage, amperes, and resistance. Two units of work measurement are watts and horsepower. To obtain watts the voltage is multiplied times the current.

$$\text{Watts} = \text{Volts} \times \text{Amps or } W = V \times A$$

Seven-hundred and forty-six watts are equal to one horsepower. Watts are measured in thousandths of watts (KW). Electrical power is sold by kilowatt-hours (KWH).

Electrical power systems have what is called a nominal voltage rating such as 120, 240, or 480. The actual voltage an electrical load (such as a motor) is supplied with may only be 118, or 223, or 456 volts. That is, electrical systems and components have a nominal, rated, and an actual operating voltage. For example, an electrical system would have a nominal voltage of 240 VAC, while a motor would be rated for 230 VAC, and the actual RMS voltage displayed by a voltmeter with the motor operating at the end of a wire, say, 1,000 feet long, may be 223 VAC.

The difference in the nominal voltage and the actual system voltage is mostly the result of pressure-voltage drop caused by the resistance of the wires of perhaps only 3 to 5 percent. Electrical systems are composed of at least one source, a protective device, a controlling device, path, and a load (such as an appliance) that does useful work. All of these components must be sized to safely do their job in an efficient manner.

Protective devices open the circuit when either the amperage or voltage deviates from normal. Insulated electrical wires provide the path for electron flow to the load. No power conversion and transmission system is 100 percent efficient; some amount of energy is lost. Most of the energy lost in electrical circuits is converted to thermal energy, which heats up the electrical wires, switches, and loads in the system.

This heat does not provide any useful work. As the amperage (A), also called *current* (I), flow increases, the amount of thermal energy lost also increases. When a current flows over, through an electrical conductor, its temperature increases. Human eyes cannot see an effect of electrical energy flow, which is the magnetic force field, the lines of flux radiating from the electrical components wires, switches, and loads.

To keep the electrons flowing only over the correct path, it is necessary that they be electrically insulated with material(s) that tend to keep the electrons on the wire (the path). A perfect electrical conductor or insulator does not exist; all types offer some amount of resistance, and all insulating materials leak some small amount of electrons.

Conductors have low resistances, and insulators offer high resistance to electron flow. While conductors move electrons and insulators tend to keep the electrons on the correct path, the magnetic field still radiates from a conductor even with the best electrical insulation. While we cannot see this magnetic field with our eyes, test instruments have been developed to measure it.

A clamp-on ammeter (or amp-meter) is placed around a single wire to measure the strength of the magnetic field and display its intensity in units of amps. A voltmeter's test leads can be touched to uninsulated electrical parts and display the electrical pressure in volts. The resistance to the flow of electrons is measured in units called ohms (shown using the Greek letter omega, Ω). Additional details about test instruments are provided in Chapter 15, Electrical Test Instruments. In summation, while electricity cannot be seen, heard, or smelled, its variables, voltage, current, and resistance can be measured using electrical test instruments.

POWER FACTOR

The term "power factor" (PF) is an expression of the relationship of the peak voltage to the peak current. Stated differently, power factor is an expression of how far out of phase the voltage and current are to each other. Figure 1–1 shows both voltage and current peaks occurring at the same time in a circuit that has a PF of one.

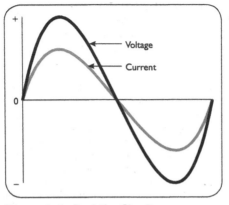

Figure 1-1 Resistive Circuit

When the voltage and current values peak at the same time, the power factor is one. When they do not, the power factor is less than one.

Many facilities have a lagging power factor of around 0.87. When the voltage changes, the current flowing in the system also changes, and the electrical system reacts to this change by creating a kind of resistance called *reactance* (X). Figure 1–2 shows the voltage and current peaks out of phase with each other. The voltage lags the current peak.

The two types of reactance are inductive reactance (X_L) and capacitive reactance (X_c). Their effect upon the power factor of a circuit is opposite to one another. Inductive reactance results in a lagging power factor. Most commercial and industrial facilities have a lagging power factor. Inductive reactance moves current flow after the flow of voltage, and capacitance moves the flow of current ahead of the flow of the voltage.

When a circuit has only resistor-type loads, such as an electrical space heater without a fan motor, the power factor of the circuit is one. When the circuit has only inductor-type loads, such as transformers and motors, the current flow peaks after the voltage peaks. That is, the system has a lagging power factor.

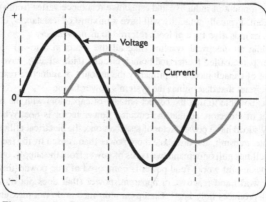

Figure 1-2 Reactive Circuit

When the circuit has only capacitors, the current flow moves ahead of the flow of the voltage, resulting in the system having a leading power factor. Figure 1-3 shows the current peak lagging behind the voltage peak in a capacitive circuit.

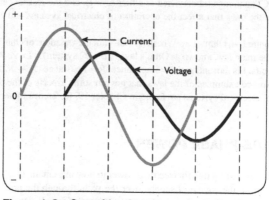

Figure 1-3 Capacitive Circuit

The amount of inductive and capacitive reactance varies from system to system. Typically a facility will have a mixture of resistance-, capacitance-, or inductive-type of loads referred to as *impedance*.

While all electrical systems (like all the roads in a town) and the smaller parts, called feeder and branch circuits (side streets), have some amount of capacitance, it is primarily the amount of inductive reactance in the circuit that determines the system's power factor.

It is possible to mix the correct amount of capacitors with the correct amount of inductors so that the circuits' power factor is one. When the facility's load has a power factor of less than one, the electrical utility must generate, transmit, and distribute more power than is used by the facility.

Facilities pull or demand two types of power from the supply, but only one does useful work. Total power is composed of true power (that does useful work) and reactive, or apparent, power (that does not do useful work—produces only heat). Clamp-on amp-meters do not measure true power, but rather they measure apparent power.

To determine the total power or true power, the installed system's power factor must be included. The result of nonproductive reactive power is that current carrying components (conductors and switchgear) must be sized larger to carry both true and reactive power.

Most of the time the primary variables of voltage, (E for AC, and V for DC) amperage (abbreviated as either A or I), and the resistance (R or Ω) are the ones that affect the operation of electrical systems and equipment.

Figure 1–4, Ohm's Law Wheel, lists all 12 of the variations of Ohm's law.

The main law, known as Ohm's law, is E (AC system) = I × Z. Where E is volts, I is current, and Z = impedance, R = resistance ($X_c + X_L$).

Worked examples of the laws that govern some aspects of the operation of electrical systems are found in Chapter 19, Reference Material.

THREE-PHASE POWER

Two wires are required for electrical power to flow in a circuit. One to pass power from the source of supply, over the path, through the protection, control, and to the load, and back to the source of supply. By changing

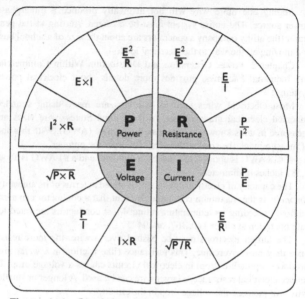

Figure 1-4 Ohm's Law Wheel

the system (starting inside of the generator), and adding one more coil of wire, the power provided can be increased by about 73 percent.

That is, by using three pulses of electrical power, each 120 electrical degrees out of phase, the amount of power produced can be increased by 73 percent. That is why most of the heavy work is done using three-phase electrical power systems. Visualizing a bicycle with three people pushing on three individual sets of peddles can be likened to the idea of a three-phase generator. Each rider provides a pulse of power (voltage) that peaks at separate times. This system produces a much smoother flow of electrical power to the loads, resulting in more power.

Adding one more wire will not magically produce a three-phase power source. The entire system must be changed, starting at the generator (the utility company's generators are about the size of a school bus) and moving all the way to the electrical load(s).

Chapter 2, Power Generation and Distribution Within Commercial and Industrial Facilities, provides more details about electrical power systems.

Most electrical wires used in buildings are made using stranded annealed electrical grade copper. In the United States, wire sizes are measured in units known as American Wire gauge (AWG). With this unit of measurement, the smaller the wire, the larger the number.

A #18 AWG is about 0.136 inches in diameter and a #1 AWG is about 0.582 inches in diameter.

The capacity of electrical conductors is stated in terms of its ampacity. *Ampacity* is the maximum continuous current that a conductor can carry without exceeding its temperature rating. Most conductors are rated for safe operation at either 140, 167, or 194 °F.

The farther electrons must be pushed over a wire, the more resistance they must overcome. This resistance (like friction in a water pipe produces a pressure drop) in electrical circuits causes a voltage drop. To reduce electrical resistance, a larger wire can be used. A longer or smaller wire will increase the voltage drop in the circuit.

Electrical conductors installed in buildings must be protected from physical damage. This is accomplished by installing them in protective envelopes called conduits, raceways, wireways, gutters, and enclosures. These envelopes are made of metal, PVC, fiberglass, aluminum, and HDPE plastic. Some are rigid, while others can be easily bent by hand.

As heat is generated when power is flowing in an electrical circuit, this heat must be transferred to the surrounding area to keep the conductors from overheating. There are several factors that cause a conductor's ampacity to be reduced, or derated. These are found in various tables in the *National Electrical Code* (NEC).

Materials such as air, paper, cotton, glass, rubber, and plastics are used as electrical insulators. Normal ambient air is commonly used as an electrical insulator. Each type of insulator is provided with a voltage rating for a specific thickness.

CIRCUIT PROTECTIVE DEVICES

Two types of overcurrent protective devices are fuses and black plastic–cased automatic switches called Molded Case Circuit Breakers (MCCB). These protective devices are designed to primarily protect insulating materials from overheating.

MCCB have two types of sensing elements, one that operates to clear running overloads, and one to clear dangerous short circuits. The overload element, called a *thermal element*, adds an amount of time delay before opening the protected circuit. The element that protects against short circuits and acts without any intentional delay is called the *magnetic element*.

When an MCCB has opened (or "tripped") to protect the circuit, it can be physically reset immediately. This allows it to be ready to protect the circuit again should the need arise.

The thermal element in an MCCB has a varying amount of time delay built into it by design. This element is typically two pieces of very thin metal bonded together. Each metal has a different expansion and contraction rate. When bonded together, with the least expansive metal (invar) on the inside, the strip of metal will bend in an arc as it is heated and cooled. This bending action allows it to be used as a trip lever to open the circuit if an overload condition is sensed by the MCCB.

Fuses are another kind of circuit protective device. When the current flowing in the circuit creates sufficient heat, the fuse element will melt. When this occurs an electrical arc will develop, accelerating the melting of the link. It is the combination of melting of the link and stretching of the electrical arc that opens the electrical circuit. Additional details about circuit protective devices are provided in Chapter 6, Personnel Protective Devices.

To improve safety many rules in the National Electrical Code (NEC, copyright © NFPA) in the United States and the International Electro Technical Commission (IEC) (standard 6036) in Switzerland are followed in over 90 percent of the world.

2

POWER GENERATION AND DISTRIBUTION WITHIN COMMERCIAL AND INDUSTRIAL FACILITIES

There are several different types of power generation and distribution systems in use today. Figure 2–1 shows how electrical power is first produced by the utility company's generators, stepped up in voltage, transmitted across vast distances, stepped down in voltage, and distributed to customers' premises wiring systems across the nation.

Figure 2–2 shows how the nation's electrical power system is being revamped to utilize more diverse sources of electrical power.

Commercial single-building electrical power systems are typically supplied by only one utility source. Utility transformers may be either cluster mounted atop timber poles, or pad mounted with underground conductors. Electrical power that radiates from a single source is called a *radial distribution system*. Figure 2–3 shows a typical radial power distribution system.

The conductors from the transformer's secondary connect to a main power distribution board in a building. This type of switchgear is known as *metal-clad switchgear*. From this point power is passed to motor control centers (MCC), local lighting panelboards, and other distribution panels. The voltages are provided at the equipment utilization level such

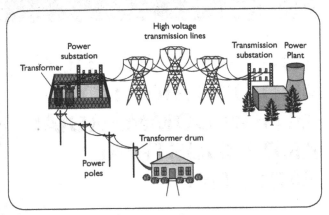

Figure 2–1 Conventional Power System

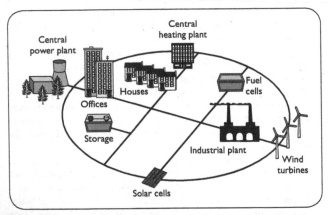

Figure 2–2 Modern Power System with Renewable Energy Sources

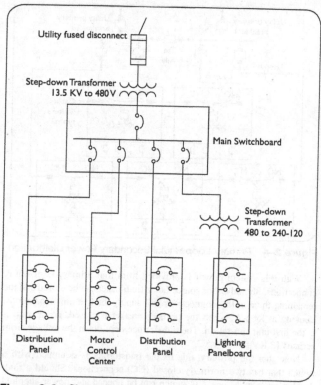

Figure 2–3 Simple Radial Electrical Power Distribution System

as 240/480 VAC. Lighting panels are typically fed from 480 to 277 step-down dry-type transformers.

Commercial, industrial, and campus-like facilities have more complex power distribution systems. Figure 2–4 shows four primary feeders and two primary loops providing power to four unit substations. This type of system offers increased reliability.

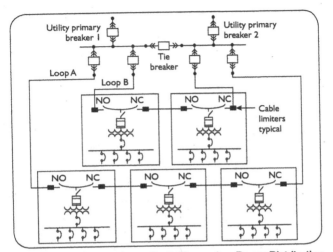

Figure 2-4 Primary Loop Radial Secondary Power Distribution

With this system power is provided from four primary-level utility connections. Should either source fail, a tie-breaker can be closed and the remaining three utility connections will supply the five unit substations. Looking at loop A, when the primary breaker is closed, power passes to the first unit substation. The transformer steps down the voltage from perhaps 13 KV to 480 VAC.

Note that the primary side of the transformer is equipped with a switch that has two normally closed (NC) connections. Should a fault develop on loop A, one NC switch can be opened and power supplied from the unit substation to the right. These switches can be moved either manually or automatically.

The secondary of this transformer feeds a bus, to which multiple individual smaller distribution boards and/or loads, such as large motors, are supplied by individual overcurrent protective devices (OCPD) such as breakers or fuses connected to the secondary bus. While loop A provides power to three unit substations, which most likely supply individual buildings or production units, loop B supplies power to two unit substations.

Failure of any one of the four primary feeder breakers will not result in long-term loss of power to any of the individual unit substations. Should both primary feeder breakers fail on loop A, both unit substations on loop B will still be provided with utility power.

Should both primary breakers on loop B fail, all three of the unit substations on loop A will still receive power. From this brief overview, it can be understood that this type of power distribution system, while more reliable than the radial-type system, is more costly to install, more complex to operate, and has more equipment that must be maintained.

Where the loads served are critical in nature, such as credit card processing centers and continuous manufacturing processes, such as oil and gas refineries and steel mills, increased reliability is a major concern. Many of these types of facilities are designed to five nines (99.999) up time or reliability.

Figure 2–5 shows a distribution system with a single utility transformer and three onsite emergency generators. During normal operating conditions, power is supplied to the facility from the utility transformer to the secondary, and then to the main power bus. Individual panelboards and loads are served at utilization voltages.

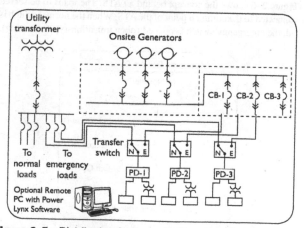

Figure 2–5 Distribution System with Three Onsite Generators

Power is passed from the bus to three individual automatic transfer switches (ATS) which, in turn, pass power to three lighting panels and isolation transformers and on to critical loads. These could be hospital operating rooms or financial data centers.

Should normal utility power be lost the computer control system would automatically start the three onsite emergency generators, cause the ATS to transfer from normal (N) to emergency power (E). Power would then be fed from the emergency-standby power bus to the three breakers, D-1, D-2, and D-3, through the ATS to EDP-1, -2, and 3, and then to LP-1, 2, and 3 and BP-1, 2, and 3. This type of system is potentially capable of island operation and possibly even black-start operation. *Island operation* is where there is no connection between the facility and the utilities' electrical supply grid.

Black-start operation is where the onsite power generation equipment is not dependent upon the local electrical utility for electrical power to initially start the generators. Notice that the onsite generators cannot provide power to all of the facility's loads, as no power is passed when utility power is offline, and the ATC switches have transferred from the N to the E positions. That is, no onsite generated power is provided to any of the "normal distribution circuits."

Figure 2–6 shows the concept behind an ATS. The loads to be served are connected to the common point of the ATS. When the normal switch is closed, the emergency switch is opened (open transition-type operation).

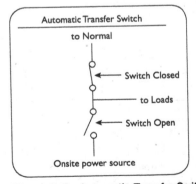

Figure 2–6   Automatic Transfer Switch

When the ATC senses a loss of utility power, a time delay begins if power has not been established by the end of this two-second delay.

Within a time of about five seconds, the generator will start and accelerate to rated speed. After about nine seconds the ATS will open the utility power connection and then close the emergency generator connection. The load will then be supplied by the emergency generator. When normal power has returned for a preset period, the ATS contacts transfer back to the normal supply, with the emergency contacts open. The generators run for a cool down time and then shut down.

Automatic transfer switches can be either open or closed during the transfer operation. That is, for a brief time both contacts can be open, for open transition operation (break before make), or both contacts can be closed for closed transition operation (make before break).

The addition of multiple onsite emergency/standby generators can provide additional savings by providing peak power, thus reducing the peak power purchased from the electric utility. When the engines are equipped for dual fuel operation, such as diesel day tanks and natural gas or propane pipeline, operation flexibility is enhanced.

ELECTRICAL POWER GENERATORS

Using motor-generators, or gen-set, to produce electrical energy, there needs to be relative motion between a magnetic field and a conductor that passes through that field. The strength of the magnetic field and arrangement and number of conductors cutting this field is what results in the production of the electrical power. With an internal combustion (IC) engine, a rotating force supplied by the engine is used to produce electrical energy. Gen-sets are commercially available from small portable models producing some 500 watts to units that produce millions of watts of power [megawatts (Mw)].

TOTAL LOADS SERVED

Typically a facility will have several classes of loads. Each class must be considered individually, summed, and then combined before the total load served can be estimated. When selecting a gen-set for an existing installation, the billing history for several years can be reviewed. A generator of sufficient size can be selected to provide the historical power used by the facility. Typically an amount between 15 to 25 percent is added for future increases in loads served. This method assumes that all loads supplied during normal operation will be supplied during emergency conditions. Another method is to develop a detailed listing of only those loads to be served during "emergency" conditions. This would be the loads actually connected to the automatic transfer switch (ATC).

TYPES OF ONSITE POWER GENERATION

Various types of occupancies have differing electrical system requirements established by authorities and national standards. There are three types of onsite power generation systems: Emergency Systems, Legally Required Standby Systems, and Optional Standby Systems.

Emergency Systems

Emergency systems are required to be installed in places of assembly, where electrical lighting provides safe exiting and panic control when the facilities are routinely occupied by large groups of people. They may also provide power to ventilation systems, fire alarm–detection systems, elevators, firewater pumps, and PA systems. In addition, emergency systems may be required in selected industrial systems, where the interruption of which would result in serious life and/or health safety hazards.

These types of systems are required by the NEC to automatically start when utility power is lost in not more than 10 seconds. A self-contained fuel supply is required to provide not less than two hours of operation at full rated demand.

Legally Required Standby Systems

The power generation systems included in this type are permanently installed at the facility. These systems, due to the nature of the facility, are required by the government authority having jurisdiction. They are required to automatically provide power to the connected loads in less than 60 seconds.

Many times these systems serve loads such as refrigeration, heating, communications, ventilation equipment, and lighting systems. The onsite fuel supply is required to be sufficient for at least two hours of full load operation.

Optional Standby Power Systems

These types of power generation systems are considered for use in public or private facilities where life safety is not a concern. These systems may provide power either automatically or manually. Loads supplied are heating, refrigeration, data processing, communications, equipment, and select industrial processes where loss of electrical power will result in only discomfort or damage to the product or process.

Interconnected Electrical Power Production Systems

In addition to the three types of conventional power systems just discussed, there is now a fourth type. This type of system may incorporate onsite bio-gas systems, cogeneration generators, emergency generators, fuel cells, flywheel systems, legally required standby systems, micro-hydro, and optional standby systems, wind turbine sources, or woodchip systems. Interconnected power production systems are hybrid systems, as they utilize multiple power sources that are not dependent upon a single energy source. As the concept of distributed generation develops and is integrated with the existing national power grid, and as diversified "green energy" sources become more economical, the interconnected electrical power production type of system will likely become increasingly common. In addition to the negative environmental impacts of burning fossil fuels, the economic efficiency of these systems is less than ideal. Consider that an onsite internal combustion engine extracts only about 36 percent of the energy contained in the fuel.

The transmission of electrical energy is accomplished at voltages higher than both subtransmission and distribution voltages. Figure 2–7 lists typical transmission, subtransmission and distribution voltages.

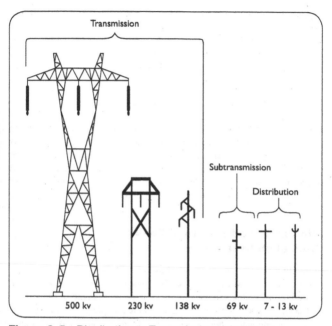

Figure 2-7 Distribution to Transmission Voltages

Electrical power is first developed at a generation plant. The voltage is then raised to the desired transmission line voltage, typically between 7 and 13 KV. The next step is to increase the voltage to between 69 and 138 KV in what is considered the subtransmission area. The next step is considered the transmission range, which starts at 230 KV to as high as 1.3 million V.

3

POWER, CONTROL, AND LIGHTING TRANSFORMER

This chapter reviews three-phase oil-filled power transformers, dry-type lighting/power, and single-phase dry-type control circuit transformers. Transformers change some aspect of an electrical energy at one frequency, voltage, phase, and current to another. Most of the time, it is only the voltage and current that are changed. Most of the electrical transformers located in commercial and industrial facilities will step down the supply voltage, resulting in the secondary current increasing for the same amount of energy. Electrical transformers are, for the most part, simply coils of insulated wire wrapped around a steel core. The three most common applications for transformers are power, lighting, and control. Because transformers are not 100 percent efficient, not all of the electrical energy put into a transformer comes out of the transformer. Most of the energy loss shows up as heat. Transformers are cooled either by an electric fan or by static air currents. Because of the increased amount of heat loss by transformers in the larger sizes, they are filled with oil to enhance cooling of the electrically insulated wires called *windings*.

In order to supply power, transformers must receive power. The power input is provided to the primary winding, and the output side, known as the secondary winding, is where the load is connected.

It is important to understand that the coils of wires on the primary side are not electrically connected to the secondary side. The primary coils are made using electrically insulated wire (magnetic wire) and are many times physically touching the coils of insulated wire on the secondary winding. While physically touching, these two coils of wire are covered with electrical insulation, so they are electrically isolated from each other. The energy transfer is accomplished magnetically. The windings are magnetically connected to each other only when a voltage is applied to the primary winding. The primary winding is one electrical circuit, and the secondary winding is another electrical circuit. Though they may be physically touching one another, both sets of wires are electrically insulated from each other. They are two separate electrical circuits that are magnetically linked together. This linkage occurs only when a voltage is applied to the primary windings which creates the magnetic field that provides the connection or links the two windings together.

This relationship may be considered somewhat like a fence between two backyards. Both have separate owners, both share a fence in common, but each owner stays on their side of the fence while talking over the fence.

When a current flows in an electrical conductor, a magnetic field radiates from the conductor (like the invisible sound of our voices). By radiating this magnetic field, a voltage is inducted into the secondary windings (like invisible sound vibrates in the inner ear).

A transformer is an electrical energy transfer device. It provides, within its limits, only the amount of energy demanded by the load connected to the secondary. It does not create this electrical energy. The electrical power must first be supplied to the transformer's primary winding, and then it is passed in a transforming manner to the secondary circuit. This is the process of stepping down the voltage and increasing the current.

Excluding the transformer's loses of about 5 percent, the power supplied to the transformer must equal the power drawn from the transformer by the load connected to the secondary windings.

Figure 3–1 shows the windings and laminated core of a transformer with the magnetic field transferring power from the primary to the secondary windings.

The transformer is supplied with AC power.

Vp = Primary voltage.

Ip = Primary current.

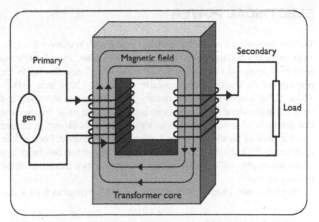

Figure 3–1 Transformer Windings and Core

Np = Number of turns in the primary winding

Ns = Numbers of turns in the secondary winding.

Is = Secondary current

Vs = Secondary voltage

When a conductor is placed within the magnetic field of an energized conductor, a current is induced into the second conductor. This process is known as *induction*. When current flows, it produces an invisible magnetic field that induces a voltage in any conductor placed within its magnetic field.

That is all the magic there is to a transformer. By varying the number of coils of wire in the primary and the secondary, the transformer can be built to step down or increase the voltage of the secondary windings. The current flow in the secondary, while provided by the primary, is controlled by the connected load. The primary only supplies what the secondary demands.

ELECTRICAL POWER

The basic electrical formula for electrical power is P (in watts) = E × I. As an example, if 1,500 watts is flowing into a transformer, about 1,500 watts must be flowing out of it. So, if 1,500 = 240 × 6.25 A is flowing in the primary and the transformer's secondary voltage is 120 V, then 1,500 = 120 × 12.5 A is flowing in the secondary. The voltage was decreased, and the current increased. This happened because of the basic requirement that power in must equal power out less losses. This change in current was the same as the change in the voltage, only in inverse. Look at it in math terms: 240/120 = 6.25/12.5. The supply voltage was two times that of the secondary, while the secondary current was two times that of the primary current.

The basic rule, known as Ohm's law, for AC is written as E = I × Z

Where E = AC voltage,
I = the current, in amps,
and Z = the impedance,
with Z being composed of $R + (X_c - X_L)$.

As the frequency did not change, and the load on the transformer did not change, then 240 = 6.25 × 38.4, and 120 = 12.5 × 9.6. The power in is equal to the power out (less losses within the transformer). That is, 240 × 6.25 = 1,500 watts of power; and 120 × 12.5 = 1,500 watts of power. While there is a lot more to transformers, the basics of how they work have been covered at this point. Losses of some 5 percent or more were omitted for clarity. Like all manmade things, transformers have limits and, when exceeded, the result is reduced service life leading to failure. The two most common causes of transformer failure are overloading and high voltage spikes.

The basic insulation level (BIL) of the transformer's windings determines the ability of the transformer to withstand high voltage spikes. The higher the BIL, the greater the voltage spike the windings' insulation can withstand without immediate failure. Most transformers are rated for a 104 °F (40 °C) ambient. When a transformer is operated in an ambient higher than 104 °F, its ability to dissipate heat is restricted. This results in an increase in the electrical insulation's temperature, and that results in a decrease in its service life.

Operating a transformer in an ambient temperature less than 104 °F can allow the transformer to carry more load without overheating. A reduction of 3.8 °F (1 °C) allows for about a 1 percent increase in transformer capacity.

INSULATION MATERIAL TEMPERATURE RISE

When a transformer is installed in a location for some time with no power applied to it, its temperature will equalize with that of the air surrounding the transformer. When power is applied to the transformer, and a load is applied to the secondary winding, the windings will begin to rise in temperature.

For example, if the ambient temperature was, say, 104 °F (40 °C), and the windings, when fully loaded, heated up an additional 131 °F (55 °C), the actual temperature of the windings would be at 203 °F (95 °C).

That is, the actual temperature of the transformer's windings, and the insulation material on those windings, will be at ambient temperature when load rise temperature equals winding operating temperature.

Insulation material is commonly listed as having a maximum temperature rise, over a given ambient temperature of 104 °F (40 °C). Sometimes this can be a bit confusing. The classes of electrical insulation used in transformers are A, B, F, and H. Class A insulation has a maximum operating temperature of 203 °F; class B, 248 °F; class F, 312.8 °F; and class H insulation has a maximum operating temperature of 374 °F.

All of these temperatures are too hot to safely touch and cannot be measured with the transformer online, so for the maintenance person on the plant floor, they are of little value.

Manufacturers do not provide an easy way to measure the temperature of the windings. Oil-filled transformers can be provided with a thermometer, allowing for the temperature of the oil to be measured, but not the hottest spot of the windings. To determine the temperature of the windings, the ambient temperature and the current draw of the secondary windings can be measured.

When the amperage is at or above the nameplate value, and the ambient is at or near 104 °F, it is likely that the windings are at their maximum operating temperature. As a quick rule of thumb: for every 1 °C below 40, the capacity can be increased by 1 percent. For every 1 °C above 40 °C, it should be decreased 1 percent.

GROUNDING

The NEC requires that all conducting non-current-carrying parts that make up the exterior of the transformer housing must be grounded. Additionally, the Code provides certain installation restrictions relative to clearances from combustible material. These details can be found in Article 450, Transformers and Transformer Vaults of the NEC.

TERMINAL CONNECTION MARKINGS

The National Electrical Manufacturers Association (NEMA) has designated terminal markings for transformers such as power, dry, and control circuit types. The primary windings of a three-phase transformer are identified by the markings H-1, H-2, and H-3. The secondary windings are marked as X-1, X-2, and X-3. Where the secondary winding has a midpoint, or center tap, it is marked as being X-2.

Where the secondary may be grounded, the X-2 terminal is to be used. On wye-connected three-phase transformers where a terminal marked X-0 is present, this terminal is to be used as the grounded, neutral conductor and is to be grounded to the grounding electrode conductor.

THREE-PHASE, OIL–COOLED POWER TRANSFORMERS

These are the large workhorses that handle massive amounts of electrical energy. These transformers are filled with oil to improve heat transfer. This type of transformer may be either pole top or pad mounted. Pole-top models have primary connections located on the top, and secondary lugs on the side, near the top of the transformer. When installed at grade, where direct contact with energized parts is possible, provisions must be made to restrict access to knowledgeable and trained personnel only. Figure 3–2 shows the internal windings of a typical pad mounted transformer with the protective outer cover removed.

Pad-mounted models are provided with two hinged doors, which are provided with a special five-pointed bolt. Typically, a padlock further restricts access to the special headed bolt. When the padlock and bolt are removed, the two access doors are opened allowing access to both the primary and secondary winding lugs. The primary and secondary conductors are routed underground to and from this type of power transformer.

Many times these transformers are supplied at utility voltages over 600 V. As these sizes of transformers are sometimes the property of the

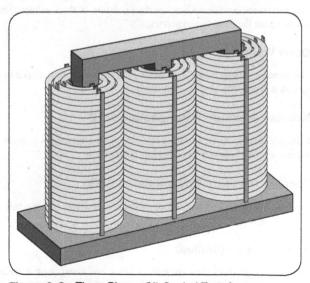

Figure 3-2 Three-Phase, Oil-Cooled Transformer

utility and sometimes the property of the facility owners, one should first determine ownership and maintenance responsibility before attempting to troubleshoot them. All work should be undertaken only after confirmation that the transformer is in an electrically safe work condition.

THREE-PHASE DRY-TYPE TRANSFORMERS

These transformers are generally available in sizes ranging from 3 to 300 KVA. While many are installed above the floor, they are generally considered to be floor mounted by design.

Primary voltages are 208, 240, 277, or 480 V. Secondary voltages are 277 for lighting-only circuits. Some provide 208 V for motor loads, while the majority provides 120 V for general outlet and lighting use. Several different methods are used in the manufacture of dry-type transformers. The differences considered here are the insulation of the windings. The following are the more common ones.

Open Wound

The standard "dip-and-bake" conductor coils are heated, then dipped in a varnish and baked in an oven until cured.

Vacuum Pressure Impregnation (VPI)

Polyester resin is applied to the coils, which then undergoes pressure-vacuum cycling followed by curing in a heated oven.

Vacuum Pressure Encapsulated (VPE)

Same as VPI process with added steps and encapsulation offering better wet-environment service.

Encapsulated (Sealed)

Standard open-wound type encased in silica and epoxy with a metal enclosure.

Cast Coil (Molded Epoxy Sealing)

These units incorporate coils encapsulated in epoxy by a molding process. They provide good service in harsh environments.

DRY-TYPE SINGLE-PHASE CONTROL CIRCUIT TRANSFORMERS

Control circuit transformers come in sizes ranging from 20 to 5,000 VA. They are available as either foot or panel-wall mountings, with open, and NEMA 1- and 3R-rated enclosures. Primary voltages available are 120, 208, 240, 277, and 480 V. Secondary voltages available are 12, 24, 120, 208, 277, and 240 VAC. The primary overcurrent protective devices (OCPD) are required by the NEC and vary depending on several factors. Article 450 of the National Electrical Code provides the requirements for protection of transformers.

4

ELECTRICAL SWITCHGEAR

The term "switchgear" refers to various combinations of bus bars, disconnects, circuit breakers, motor starters, and transformers used to control and isolate electrical equipment so that maintenance and repairs can be completed safely.

In North America, NEMA, IEEE, and ANSI standards are used to establish designs, ratings, and specifications. In other nations, IEC standards with local national variations are used.

MOTOR CONTROL CENTER

A motor control center (MCC) is a factory-made assembly of electrical devices that control the operation of typically three-phase motors. In the United States, MCCs are contained in sheet metal enclosures of modular design. MCCs can include fused and nonfused three-phase disconnects, control circuit transformers, motor starters, pilot control devices, and power monitoring/recording/management systems. Figure 4–1 shows a typical modular MCC. The modular nature of MCCs allows for many variations. While available in both low and medium voltage designs, most are of low voltage (600 VAC).

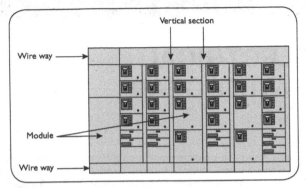

Figure 4–1 Motor Control Center

PRIMARY FUNCTIONS OF SWITCHGEAR

MCCs, panelboards, load centers, and distribution boards facilitate the supply of the electrical power to the various loads within a facility. Many times this distribution is both to individual loads and smaller distribution boards, which in turn supply power to additional loads and/or to still smaller distribution boards.

The primary functions of electrical switchgear are to provide:

▦ Protection and coordination
▦ Isolation from energized parts
▦ Local and remote switching control

Switchgear Components

The following components are many times used to make up electrical switchgear:

▦ Fused and nonfused disconnects
▦ Contactors and motor starters
▦ PT and CT transformers

- Circuit breakers and fuses
- Power monitoring equipment
- Alarms
- Anticondensate cubical heaters

Additional devices such as control, energy monitoring, surge protection, and over/ under-voltage protective relays may also be provided.

LOW-VOLTAGE POWER DISTRIBUTION

Electrical power is typically routed to a single point. From this one point, power is distributed through the building. The following is a listing of the various distribution boards used in low-voltage power systems:

- Main switchboard
- Motor control center
- Subdistribution boards (panelboards)
- Final distribution boards (for lighting and power, called *load centers*)

The physical locations of the various types of switchgear vary with the nature of the facility. Some are near the main switchboard, while others may be near the loads served, with other distribution boards located throughout the facility. Today most switchboards are built in a modular fashion. This allows for popular voltage, phase, and amperage sizes to be built as a type, and then various individual components selected and installed into the board as needed by the specific building. The various components can be one of three types, bolted units that cannot be isolated from the supply. These must be worked hot, or the power to the entire facility must be killed (placed in an electrically safe work condition).

An improvement from a safety point is the use of disconnectable types. With this type each individual unit is mounted on a module that can be removed from the board once the load has been turned off. This is made possible by the use of male and female separable connections between the individual bus bars and the module. The final option is the use of drawer-type draw-out units that can be slid out of the board. Once the draw-out unit has been removed, it can be repaired or replaced. To improve safety, modern modular switchgear is built so that there are various isolating metal barriers between the supply bus bars, the individual modules, and between the modules and each other.

BUS BAR AND CIRCUIT BREAKER ARRANGEMENTS

NEMA switchgear has the breakers in a three-phase panel arranged in three vertical columns. Figure 4–2 shows the various arrangements of bus bars in MCCs. Each column connects the breakers in that column to one bus bar. The breakers are numbered from left to right, and from top to bottom. When the bus bars are arranged vertically, the first bar is connected to phase A. The next bar (center) is connected to phase B, and the third bus bar connects to phase C. When the bus bars are arranged horizontally, the top bus bar is phase A, the next lower bus bar is phase B, and the lowest bus bar is connected to phase C. The individual three-phase breakers connect to each of the three bus bars.

When the panel is a single-phase model, the bus bars are typically arranged vertically. The first breaker is installed in the first slot on the left and connects to phase A. The first breaker on the top right connects to phase B. The first breaker is numbered one, on the top left, followed by two on the top right, the next lower breaker on the left is numbered three, and the next lower breaker on the right is numbered four. This results in all of the odd-numbered breakers being installed on the left, and all of the even-numbered breakers being installed on the right, in a top to bottom order. Figure 4–3 shows a typical bus bar arrangement in an electrical panel.

NEMA panels have their main power lugs located at the top (when installed vertically) of the panel. Some panels are provided as main lugs only, while others have a main circuit breaker. When the panel has a main circuit breaker, it is installed at the top of the panel and connects to the bus bars. While not common, it is possible to have a circuit breaker installed in a back feed position. When this is done, a breaker installed in a location typically used for a branch circuit breaker is used to feed power to

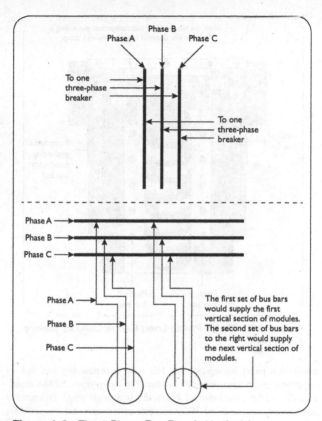

Figure 4–2 Three-Phase Bus Bars in Vertical Arrangement

the panel from the normal load terminals of the breaker. The NEC places restrictions on the use of back feed circuit breakers.

When panels are designed as service equipment, the neutral and ground bus bars are provided with a means of connecting the ground bus bar to the panel's metal enclosure. This results in a single-phase

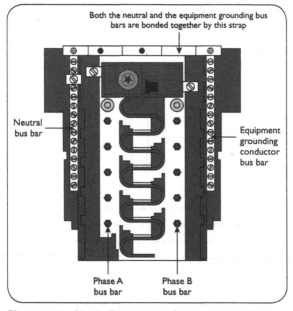

Figure 4–3 Single-Phase Load Center Used as Service Equipment

distribution panel having two hot bus bars, a neutral bus bar, and an equipment grounding conductor bus bar. For many years, NEMA standards allowed for a maximum of 42 circuits in a single panel. This restriction has now been removed. IEC consumer units (somewhat like a NEMA load center with miniature circuit breakers) may have between 6 and 24 breaker slots.

When the panel is used as service equipment, a means is provided to connect both the neutral and the ground bus bars together. Caution should be exercised when the panel is not used as a service panel. The ground and neutral bus bars must not be bonded together. That is, they must be electrically isolated from each other. The ground bus must be bonded to the enclosure in all cases. Many times this is done with a green-headed threaded machine screw.

Circuit Breakers

A circuit breaker is a special type of switchgear designed to be able to interrupt fault currents without damage to itself. Their construction allows them to interrupt fault currents of many hundreds or thousands of amps.

The methods used in quenching the arc falls into four types:

■ Oil circuit breakers rely on vaporization of some of the oil to blast a jet of oil through the arc.
■ Gas-insulated (SF6) circuit breakers sometimes stretch the arc using a magnetic field, and then rely on the dielectric strength of the insulating gas to quench the stretched arc.
■ Vacuum circuit breakers have minimal arcing (as there is no gas to ionize), so the arc quenches when it is stretched a few thousandths of an inch. Vacuum circuit breakers are frequently used in medium voltage switchgear to 35 kV.
■ Air circuit breakers (ACB) use either ambient or compressed air to extinguish the arc.

Circuit breakers are designed to terminate current flow between three to five cycles.

OPTIONAL PROTECTIVE DEVICES

Besides the standard thermal-magnetic and magnetic only circuit breakers (MCP), there are many other protective devices that may be added to a lineup of switchgear. The following is a brief listing:

■ Over- and/or under-voltage
■ Phase sequence
■ Thermal overload
■ Instantaneous

- ▓ Phase loss
- ▓ Direction overcurrent
- ▓ Physical lock-out devices
- ▓ Differential protective relay

METAL-CLAD SWITCHGEAR

Metal-clad switchgear is defined by ANSI/IEE C37.3202 as metal-enclosed power switchgear. This equipment has removable, draw-out-type switching and interrupting devices with connect and disconnect positions. The major parts of the switchgear are enclosed by metal barriers that are grounded for compartmentalization. When the switchgear is removed, sometimes automatic shutters cover primary and secondary stabs. All primary bus conductors and connections are typically covered with insulation material. Exercise caution when working in switchgear as this insulation material is not intended to protect personnel from electrical shock.

Metal-Clad Switchgear Voltage Ratings

Some metal-clad switchgear manufacturers offer 38-kV-class switchgear with a 170-kV basic insulation level (BIL). SF-6 insulated (GIS) switchgear may have even higher BILs.

CONTINUOUS CURRENT RATINGS

Current ratings of bus bars in the metal-clad switchgear are 1,200 A; 2,000 A; and 3,000 A. Some manufacturers can supply 4,000 A continuously rated bus bars.

Breaker continuous current ratings are normally 1,200 A; 2,000 A; and 3,000 A for 15 kV class. Some manufacturers can increase these breaker ratings to as high as 4,000 A. This is accomplished by adding cooling fans to the breaker itself or by adding the cooling fans to the cell, allowing a standard 3,000 A breaker providing a total continuous breaker cell rating of 4,000 A. With 27 and 38 kV class, breaker ratings are 1,200 A and 2,000 A.

ARC-RESISTANT SWITCHGEAR

This class of product has gained popularity in North America over the past few years and in particular in Canada where it is said to have originated.

The applicable North American Standard used for testing is EEMAC (Electrical Equipment Manufacturers Association of Canada) G14–1 Procedure for Testing the Resistance of Metal-Clad Switchgear Under Conditions of Arcing Due to An Internal Fault.

In the last few years, a few manufacturers have been marketing "arc resistant" switchgear. This equipment provides an increased level of safety to personnel who may be in close proximity to the equipment should an arc fault condition develop within the switchgear.

Arc-resistant switchgear test criteria to the EEMAC standard includes:

- ■ Properly secured doors, covers, and so forth, that do not open under tested arc fault conditions
- ■ No hazard due to flying parts
- ■ Arcing does not cause holes in sides of switchgear covers
- ■ Flame test indicators
- ■ Grounding connection remains effective

Arc-resistant switchgear is now certifiable by UL and UL-C (for Canada). Some metal-clad switchgear manufacturers have tested their arc-resistant switchgear to what some consider to be as rigorous IEC standards.

EEMAC-tested arc-resistant switchgear includes three basic types of accessibility:

- ■ Accessibility Type, A which is arc-resistant construction at the front of the equipment only.

▪ Accessibility Type B, which is arc-resistant construction at the front, back, and sides of the equipment.

▪ Accessibility Type C, which is arc-resistant construction at the front, back, and sides and between compartments within the same cell or adjacent cells.

There is an exception to Type C allowed in which a fault in the bus bar compartment of a feeder cell is allowed to break into the bus bar compartment of an adjacent feeder. Some manufacturers have undertaken tests on metal-clad switchgear to preclude this from taking place and, instead, contain the arc within the bus compartment of any feeder cell.

Some overseas manufacturers offer switchgear with spring dampers that open to relieve the pressure during an arc blast fault.

5

ELECTRICAL CIRCUIT PROTECTIVE DEVICES

Overcurrent protective devices (OCPD) can be either one-time use fuses or resettable circuit breakers. Circuit breakers come in the following types:

- Molded case circuit breakers
- Miniature circuit breakers
- Vacuum circuit breakers
- Insulated case circuit breakers
- Low-voltage power circuit breakers
- Supplemental protectors
- Motor circuit protectors
- Oil circuit breakers
- Gas-insulated circuit breakers

Figure 5–1 provides a typical diagram of the relationship between time and current for inverse time circuit breakers. The words inverse time are used to communicate that as the current increases, the time required for the circuit breaker to open decreases. Circuit breakers operate to protect the circuit from abnormal and potentially unsafe conditions. Some circuit breakers have both a thermal and a magnetic sensing element. The thermal element responds to running overloads (percentage increases above rated current), while the magnetic element responds to short circuits (multiples of rated current).

Inductive-type loads such as motors, when power is first applied, pull multiples of their rated current. This is normal; there is nothing wrong with the motor, so the OCPD should not open the circuit. If the motor was running and shorted out and started pulling, say, 300 amps, the OCPD

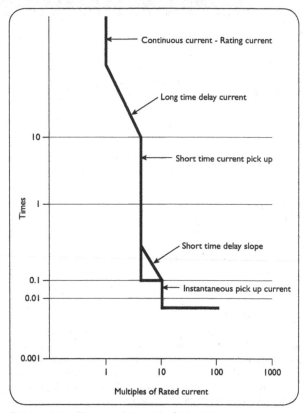

Figure 5–1 Thermal-Magnetic Circuit Breaker Inverse Time Current Curve

should open the circuit immediately without delay under this short circuit condition. In the preceding, two different cases have been considered, one where the OCPD should act instantly, without delay, and another where it should delay, opening the protected circuit. As it takes time for the thermal element to heat up, it protects the circuit from running overloads. A magnetic field can build up instantly, so the magnetic element acts to protect the circuit without delay against short circuits.

MOLDED CASE CIRCUIT BREAKERS

Molded case circuit breakers (MCCB) are by far the most commonly used breakers in residential, commercial, and industrial facilities. MCCBs have current ratings in standard increments between 15 and 6,000 A. Figure 5–2 is a drawing of a typical three-phase MCCB. Internally, they may have either a thermal or magnetic trip element, or they may have both types of trip elements.

Both miniature and molded case circuit breakers have contacts exposed to ambient air. They are either electromechanically controlled or microprocessor controlled and are often used for main power distribution in small and medium-sized facilities and industrial plants.

An MCCB that has only a magnetic element is called a motor circuit protector (MCP). MCPs are used to provide short circuit protection, while a motor starter provides the running overload protection.

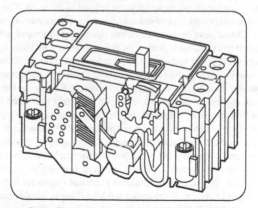

Figure 5–2   Three-Phase Molded Case Circuit Breaker

In low-voltage systems circuit breakers come in one-pole for 120 and 277 VAC, two-pole for 208, 240, or 480 VAC, and three-pole for 208, 240, and 480 VAC. When the neutral conductor must be opened, four pole breakers are used. Multipole breakers open all poles (switches) at the same time. A breaker must not be installed where its voltage rating will be exceeded. For example, a 120/240 slash-rated breaker should not be used to control a 277 V lighting circuit.

Circuit breakers also have an ampere rating. Breakers under about 200 amps are tested in free air. When they are placed in a metal box such as a load center or panelboard, they can carry continuously only 80 percent of their rating. Breakers rated above about 200 amps are 100 percent rated. Therefore, a 30-amp breaker can only carry 80 percent of its rating continuously.

$$80\% \text{ of } 30 \text{ is } (.8 \times 3) \text{ 24 amps}$$

When the breaker is located in a much hotter location, say, in a poorly ventilated boiler room, the thermal element may trip the breaker sooner than it normally would. Where this occurs, an ambient compensated breaker should be considered for use.

The operating handle of a breaker can be in either the on or off position, or a little over the midpoint, which is the tripped position. MCCBs can be mounted by either some type of stab-lock, push-in/pull-out method, or bolted in place. Power must be removed before attempting to remove a breaker. With bolted breakers, check to see if the bolt is still hot, most likely it may be necessary to kill power to the entire panel before the bolt(s) can be removed.

Breakers with a thermal element exhibit an inverse current and time relationship. That is, as the amount of current flowing in the circuit increases, the sooner the breaker will open to protect the circuit component from overheating. Breakers can be considered as being high temperature limit switches. When the temperature of the conductor and its electrical insulation increases, the breaker's thermal element temperature increases, and when the point is reached that the insulation is in danger of being damaged, the breaker opens its contacts, stopping the current flow in the circuit. The size of an insulated electrical conductor is typically selected for 125 percent of the connected load. That is, breakers are sized to prevent the circuit from overheating.

As an example, for an appliance with a rated load of 16 amps, the circuit conductor would be rated at 20 amps, and the circuit breaker at 20 amps. Should the circuit begin to pull 20 amps for an extended time, the breaker's thermal element would open the circuit. Should this same appliance short out internally and begin to pull short circuit amps of over 100 amps, the magnetic element would open the circuit without delay. When a breaker trips out, the operating handle moves to a position about 60 percent of the way between the on and off positions. As a safety feature, when a breaker opens it means that something is wrong and should be corrected.

MINIATURE CIRCUIT BREAKERS

Miniature circuit breakers (MCBs) typically have a rating of less than 200 A and have either thermal, or thermal and magnetic elements. Mounting options include flush, surface, or use of a 35 mm DIN rail. Connection options include lugs on both ends, bolt on, or plug in. Some are rated for switch duty (SWD).

Many are Heating, Air Conditioning, and Refrigeration (HACR)–rated and are also rated for use with high intensity discharge (HID) lighting. Miniature circuit breakers are used extensively in Europe in residential panelboards known as consumer units (CU), and in commercial distribution boards. MCBs should not be confused with a supplemental protector used extensively in the United States, as they physically look very much alike.

Voltage ratings include 120/240 VAC, 125 VDC, and 277 VAC.

VACUUM CIRCUIT BREAKERS

Figure 5–3 shows the electrical contacts enclosed in a vacuum tight enclosure. Vacuum circuit breakers (VCBs) have contacts encased in a vacuum bottle. A vacuum is free of any gas (air) to support an electrical arc. The arcing column is made up of metal vapor and electrons coming from the switch contacts. A very small amount of contact material is vaporized and the arc develops in this metal atmosphere, which fills the space. When current

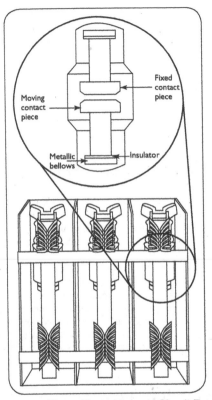

Figure 5–3 Three-Phase Vacuum Circuit Breaker

flow decreases, these metal vapors condense on the electrodes and a metal screen. Vacuum circuit breakers have current ratings of up to 3,000 A and 35 kV. Vacuum breakers have longer service lives than air circuit breakers. In the medium-voltage applications, they provide an increase in reliability and capacity over MCCBs and insulated case breakers. With contacts contained in an airtight chamber, there is no risk of explosion, fire, or external effects arising from the vacuum bottles during the breaking process. They have longer maintenance intervals than other types of breakers.

INSULATED CASE CIRCUIT BREAKERS

Insulated case circuit breakers (ICCBs) are available in frame sizes from 400 to 5,000 A, with interrupting ratings of 65, 85, and 100 kAIC (thousand ampere interrupting capacity). Short time ratings of 25, 35, and 65 kA are available from several manufacturers. These breakers are rebuildable. They typically are made in a draw-out arrangement.

LOW-VOLTAGE POWER CIRCUIT BREAKERS

Low-voltage power circuit breakers (LVPCBs) are available in sizes ranging from 800 to 5,000 A, with interrupting ratings of 65, 85, and 100 kA at 600 V. Short time ratings of 35, 65, 85, and 100 kAIC are available.

SUPPLEMENTAL PROTECTORS

Supplemental protectors devices (SPD) are molded case overcurrent devices listed for use within an machine to provide overcurrent protection for only the components located downstream of the SPD. While they physically look very like a MCCB, they are not allowed to provide branch circuit protection.

MOTOR CIRCUIT PROTECTORS

Motor Circuit Protector devices (MCPDs) are MCCBs with only a short circuit protecting element. While they physically look very much like a MCCB, they do not provide overload protection. They are used in conjunction with motor starters which provide the necessary running overload protection.

OIL CIRCUIT BREAKERS

Oil circuit breakers have all but been eliminated from medium voltage applications due to the inherent fire danger. Oil circuit breakers are still in use in some large medium and high voltage applications, but are being replaced by vacuum circuit breakers and insulated gas (SF-6) circuit breakers.

GAS INSULATED CIRCUIT BREAKERS

A man made synthetic gas known as SF-6 is used in larger circuit breakers to provide an environment for high voltage and amperage contacts in medium and high voltage applications. SF-6 provides better electrical arc extinguishing abilities than both air and vacuum type circuit breakers. Virtually all high voltage circuit breakers in utility and industrial applications have their power contacts enclosed in a pressurized SF-6 electrical insulating gas environment.

FUSES

Fuses are commonly made using a lead alloy metal link enclosed within an environmentally sealed and electrically insulated tube. Their operation is dependent on the metal link melting to open the electrical circuit. Figure 5–4 shows both barrel and blade type fuses and the internal link.

Fuses are used frequently for circuit protection and work well, yet their design requires that they must be replaced every time they operate to clear an overcurrent condition.

Fuse element deterioration can occur due to environmental and physical stresses imposed by repeated short duration electrical overloads that do not cause the link to melt. As they are sealed in a nontransparent tube, maintenance personnel cannot determine this deteriorated condition. Fuses are very economical and available in a vast range of sizes in

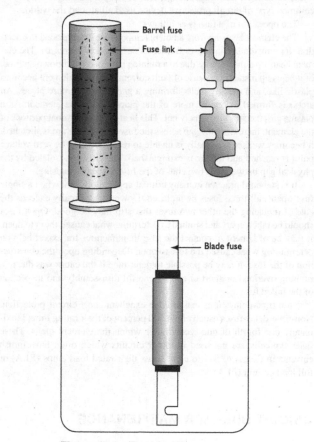

Figure 5–4 Electrical Fuse

terms of voltage, amperage ratings, and interrupting ratings. They are a common type of circuit protective device used all around the world.

The operation of a fuse is as follows:

The element heats up from ambient temperature when placed in operation. Its temperature increases further as current flow increases. The element heats up either slowly, due to a running overload condition over time, or it heats up instantly because of fault current flow. The element becomes plastic-like, and quickly melts, forming a gap in one or more places. An arc(s) is formed in one or more of the places along the element. As a plasma gas, the arc is extremely hot. This heat melts additional portions of the element, increasing the gap across the fuse link. As the arc is stretched, it becomes weaker and finally is unable to restrike when the zero voltage point is reached and the arc is extinguished. The circuit is isolated by the physical gap between the two ends of the fuse assembly housing.

It is standard practice in many critical applications that when a single fuse opens, all three fuses be replaced. Downtime typically exceeds the cost of replacing the other two fuses that still may be good. Open fuses should be taken apart and studied to determine what caused them to open. It may be of benefit to contact the fuse manufacturer for expert help on determining what caused the fuse to open. Depending upon the construction of the fuse, it may be possible to determine if the cause was due to a running overload or short circuit by careful disassembly and inspection of the failed fuse.

Non-time-delay fuses can provide excellent short circuit protection. Non-time-delay fuses usually hold 500 percent of their rating for approximately one-fourth of one second after which the element melts. These fuses typically are not used in motor circuits which often have inrush currents in excess of five to eight times their rated load amps (RLA) or full load current (FLA).

CIRCUIT BREAKER MAINTENANCE

MCCBs are contained within a case that is held together with rivets, so there are no components that can be field maintained in sizes up to about 200 amps. Breakers over the size of about 200 amps are designed so that they can be rebuilt. Insulated case, low, and medium voltage power circuit breakers (ICCB, LVPCB, and MVPCB, respectively), and vacuum circuit breakers (VCB) can be rebuilt. All breakers should be inspected and tested periodically. Most manufacturers specify that a breaker be manually turned on and off about six times periodically. This aids in redistributing the lubricant inside of the breaker.

A noncontact infrared thermometer should be used to compare the temperature of the individual termination points to each other, and differences of three degrees or more should be investigated.

The voltage from phase to phase should also be checked on the load side of the breaker. Differences of 3 percent or more should be investigated.

Breakers can explode, just like a bomb! Before operating any breaker, the front panel cover must be fully secured in place before they are operated. To reduce the potential for injury, the "step to the side" method is recommended. Figure 5–5 shows the recommended step to the side position. It requires one to place a leather-gloved hand on the breaker's operating handle, take a step to one side of the panel, and then operate the breaker.

Should an extremely bad short circuit occur, the leather glove would provide some protection for the hand from being burned. By standing to the side, any flames, sparks, or missiles will not impact the face or torso.

Details as to the specific maintenance task for circuit breakers can be found in various standards produced by The International Electrical Testing Association (NETA) and the Recommended Equipment Maintenance Standard 70-B developed by the National Fire Protection Association (NFPA).

COORDINATION BETWEEN PROTECTIVE DEVICES

When the various circuit protection devices are properly coordinated, only the smallest portion of the distribution system should experience an outage. Selective coordination is an engineering task. Maintenance and

Figure 5–5 Safety Side Step

calibration of circuit breakers are maintenance tasks. Poor engineering or maintenance can result in a lack of coordination between overcurrent protective devices.

6

PERSONNEL PROTECTIVE DEVICES

INTRODUCTION

Generally, it can be stated that a circuit breaker is intended to protect the electrical conductors in the circuit from overheating. It can also be stated that the equipment grounding (bonding) conductor (EGC) is required by the NEC to be connected so that a low impedance (not defined) fault current path is provided for the main purpose of removing the voltage from metal parts that are subject to being contacted (secondary-indirect contact) by people.

In other words, breakers protect the circuit conductors, and the equipment-grounding conductor (for a grounded system) protects personnel from indirect–fault-abnormal type contact with current-carrying conductors. This is accomplished by sensing a level of current less than what the human body can safely withstand. People do not normally make indirect, unintended, or accidental contact with current-carrying components. Circuit protective devices such as fuses and circuit breakers will not keep a person from getting shocked, hurt, or even killed by direct intentional contact. In fact, people are killed by electricity almost daily by systems that have good fuses and circuit breakers.

There is a family of protective devices that can be called "personnel protective devices," indicating that their primary purpose is to protect personnel, not electrical conductors nor equipment. These devices are known by an "alphabet soup" listing, such as AFCI (for arc fault circuit interrupters), GFCI, (for ground fault circuit interrupters), and so on. While there are some exceptions, generally by design, these devices operate when they sense a very small level of current imbalance or flow where it should not be present.

Some of these devices (AFCIs) may look at the waveform of the current passing through the device. Devices that sense a current imbalance are known as *residual current devices* (RCDs) and *ground fault circuits* (GFCs). This family of devices, by design, has trip points well below those that can cause harm to human beings. That is, they trip when the current sensed is in the milliamp (one-thousandth of one ampere) range. These devices operate on a current much smaller than a circuit breaker will. They open the protected circuit between 25 and 40 milliseconds. That is, in less time than the onset of a medical condition known as *ventricular fibrillation.* This is an irregularity of the heartbeat in which the fibers of the heart muscles work without overall coordination, resulting in failure of the heart to pump blood. If the condition is significant, as is the case when one is electrically shocked, the terminal result is loss of life.

Some members of this family of devices may have setpoints well above those that can provide protection of people [4 to 6 milliamperes (mA)], that is, setting in the area above 30 mA to as high as 500 mA(one-half of one ampere). An RCD that is intended to protect an appliance power cord will typically have a setting of some 30 mA. Some of these devices can offer protection even when an equipment grounding conductor is not present, as they operate not on current flow over the EGC, but upon a current imbalance between the two normal current-carrying conductors of the circuit, such as the hot and the neutral.

Devices with setpoints in the hundreds of milliamperes provide fire protection by limiting the energy flow to below that required to start a fire under specified test conditions. These devices may automatically close (nonlatching) when power is provided to the line side terminals, while others require the device's reset button be manually depressed (latching). While the most common types of RCDs are outlet or 120-V single-phase duplex receptacles, four-wire, three-phase models are available.

Some of these devices can have downstream loads connected to them in such a manner as to provide RCD protection to other power outlets. The following is a listing of a few members of this family of personnel protective devices:

▦　Appliance leakage current interrupter (ALCA)
▦　Arc fault circuit interrupters (AFCI)

- Equipment leakage circuit interrupter (ELCI)
- Ground fault circuit interrupters (GFCI)
- Ground fault protective equipment (GFPE)
- Immersion detection circuit
- Interrupters (IDCI)
- Leakage current detection interrupters (LCDI)
- Residual current devices (RCD)

A definition of a ground fault circuit interrupter is "a general use device whose function is to interrupt the electric circuit to a load within an established period." There is a class A GFCI that trips when the ground fault current exceeds 5 mA and there is a class B GFCI that trips when the ground fault current exceeds 20 mA. A class B GFCI with a 20-mA trip level is to be used only for protection of underwater swimming pool lighting fixtures installed before adoption of the 1965 National Electrical Code (NEC). When using a class B GFCI, the swimming pool lighting circuit must be disconnected before servicing or relamping the lighting fixture.

Two sections of Figure 6–1 are of the most interest. First is the typical plot of current and time for a GFCI. The next is the plot of the onset of

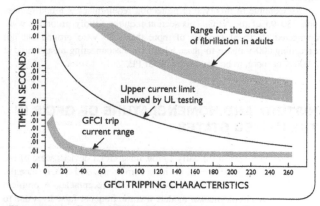

Figure 6–1 Typical GFCI Tripping Characteristics

heart fibrillation for an adult. As can be seen in the figure these devices provide ample safety room for an adult in normal health.

GROUND FAULT PROTECTION OF EQUIPMENT

Section 426–28 of the NEC requires ground fault protection of equipment for fixed outdoor electric deicing and snow-melting equipment and Section 427–22 of the NEC requires the same for electric heat tracing and heating panels. The trip ratings for these devices are usually in the 30-mA range and are not intended as personal-protection devices. Ground fault sensing and relaying equipment is intended for use in electrical power distribution systems rated at a maximum of 600 V.

GFPE devices operate to cause a disconnecting device to open at a predetermined minimum value of ground fault in accordance with the NEC. Ground fault protection of equipment is defined in the NEC in Article 100 as "a system intended to provide protection of equipment from damaging line to ground fault currents by operating to cause a disconnecting means to open all ungrounded conductors of the faulted circuit."

An example of this type of protection requirement is found in Section 230–95 of the NEC. This section requires solidly grounded wye-connected electrical services of more than 150 V to ground but not exceeding 600 V phase-to-phase with main disconnecting means rated at 1,000 A or more, to be provided with GFPE.

HISTORY AND NOMENCLATURE OF GFCIS IN THE UNITED STATES

In the United States in the early 1970s, most GFCI devices were of the circuit breaker type. However, the most commonly used GFCI, since the early 1980s, are those that are built into outlet receptacles. A problem with the early circuit breaker models was the frequent false trips due to the poor AC characteristics of 120-V insulations, especially in circuits having longer cable lengths. With long cables sufficient current leaked along the length of the conductors' insulation that the breaker might trip with the slightest increase of current imbalance.

TROUBLESHOOTING GFCIs

Figure 6–2 provides a diagram of the major components of a modern GFCI. As an aid to troubleshooting GFCIs, let us begin by reflecting on the not-so-perfect history of GFCIs.

When first delivered to the U.S. market, these devices quickly earned a bad reputation for tripping when no fault was present. This was especially true on construction sites where the leakage from long and multiple extension cords contributed to the tripping of the GFCI. The next trouble to beset the early models was when they were subjected to a high voltage spike. A review of these types of failures resulted in manufacturers and UL allowing the design to be changed so that a higher milliamperage level would be allowed. The UL listing procedure also allowed the device to fail when closed.

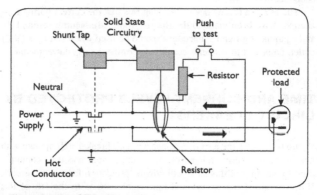

Figure 6–2 GFCI Circuit Diagram

That is, when the electronic components within the GFCI were damaged by a high voltage spike, the push-to-test and reset buttons would continue to move, but the power to the load would not be cut off when the test button was pressed. Consumers and many electricians incorrectly assumed that when the push-to-test and reset buttons where moving, the GFCI was operating properly.

Thankfully several years later, manufacturers and UL again changed the test protocol to require the device to fail open when the electronic components were damaged.

When testing a breaker equipped with a GFCI function, a load such as a radio can be connected to the device to confirm that power is removed when the test button is depressed; when the sound goes off that is proof that the GFCI tripped.

This test does not provide information as to the value of current at which the device operated. There are test instruments that inject adjustable levels of ground fault current that should be used to confirm that a GFCI is operating in the correct milliamperage range.

When GFCIs first came to the U.S. market, they could be installed incorrectly and would still pass power while not providing the GFCI function. Sometimes, they were installed with the line and load connections reversed. UL now requires that the device not operate when it has been incorrectly wired.

The use of a testing device that confirms that the correct wiring connections have been made to the Hot (ungrounded), Neutral (grounded), and equipment ground connections should be used when testing a GFCI outlet. Some test instruments can also determine the trip current value.

TIME AND CURRENT LEVELS PROTECTED BY CIRCUIT BREAKERS

Figure 6–3 shows the area over which a circuit breaker will operate with normal manufacturing tolerances. All of the current and time combinations to the right of this band will provide protection from overcurrents. However, all of the current and time combinations to the left of this band are not protected by the circuit breaker. The area to the left of the

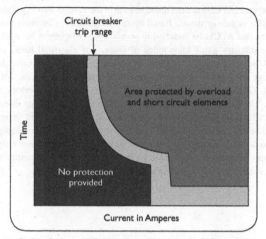

Figure 6–3 Areas of Protection Provided by Circuit Breakers

manufacturer's tolerance band is the area where some other type of personnel protective device must provide protection. This is the area where RDCs, or GFCIs, provide protection to either personnel or equipment. As further research is conducted low current levels are being shown to be capable of providing sufficient energy to start fires.

ARC FAULT CIRCUIT INTERRUPTERS

By utilizing a solid-state microprocessor and modern algorithms, a new family of protective devices has been created. These protective devices provide protection from nonstandard current flows. Common devices such as ballast, universal motors, and relays develop current waveforms that are safe. Loose connections, series, and parallel arcing faults are not safe and are not normal. Circuit breakers and GFCIs do not operate to protect the circuit when arc fault conditions occur. Arc fault circuit interrupters (AFCIs) are designed to look at the current flowing and determine if it is safe or unsafe based upon an algorithm. Currently the NEC requires that AFCIs be installed in several types of circuits in residences. As the industry gains knowledge of sources of electrical fires, and so-called "smart homes," additional circuit and personnel protective devices utilizing microprocessors will become commonplace. Fuses, MCCBs, RCDs, GFCIs, and AFCIs are not all of the protective devices in use today. The future of circuit and personnel protection will be filled with more advanced microprocessor-equipped devices. The next generation of protective devices will continue to provide protection for equipment and personnel. These next-generation devices will provide not only evolutionary protection enhancements, but also control, monitoring, scheduling, trending, and alarm functions as well.

7

THREE-PHASE MOTORS

About 90 percent of the important motors in commercial and industrial facilities are three-phase induction motors, which have three important parts: housing, stator winding, and the rotor with its shaft and the bearings that support it. Most motors made in the United States are made according to standards developed by the National Electrical Manufacturers Association (NEMA).

Motors intended for use in countries such as Europe are manufactured according to standards developed by the International Electro Technical Commission (IEC). Only NEMA-designed motors will be reviewed.

Many, but not all three-phase induction motors are designed to be operated on either 240 or 480 VAC. Motors have a service factor (SF), of either 1, 1.15, or 1.25.

A motor's SF indicates the percentage of running overload that it can withstand without damage. This extra capacity is based upon ideal testing lab conditions. Seldom will a motor be operated under ideal conditions. For example, if the voltage is a little bit below the motor's rated voltage, some of that service factor will be used.

Several studies indicate that many motors are not loaded to their rated capacity. When this is the case, the motor will operate less efficiently. In other words, the efficiency of a motor is a function of the load and its design.

When a motor fails, it may be tempting to replace it with a larger and more efficient motor. However, this may not be the best decision. As an example, a 5 horsepower (HP) motor failed and a 7 ½ HP motor is used to replace it. It is most likely that the 5 HP motor was not fully loaded. The 7 ½ HP motor may not even be loaded to 65 percent of its capacity.

This lightly loaded motor will operate far below its rated capacity and efficiency. The root cause of the motor failure should be determined and corrected.

ELECTRIC MOTOR PRINCIPLES OF OPERATION

An electric motor is an electromagnetic mechanical device that changes electrical energy into the spinning mechanical force of its output shaft. It derives its name from the fact that the moving part, the rotor, is not electrically connected to the power supply.

The electrical currents that circulate in the rotor's conducting bars are not directly produced by the voltage supply to the motor.

These currents are the result of the voltage being induced in the rotor by the magnetic field of the motor's stationary coils of wire called the *stator winding*. When the rotor is subjected to these high-intensity, rotating magnetic fields, strong magnetic fields of equal and opposite polarity are induced into the rotor. These two sets of magnetic fields interact with each other to cause the rotor to rotate. This interaction results in the development of torque at the motor's output shaft. This torque causes the rotor to accelerate in an attempt to reach the same speed as the rotating magnetic field of the stator windings.

Decreasing Torque and Current

As the rotor begins to rotate at a speed closer and closer to that of the stator's magnetic field, the amount of torque produced by the rotor begins to fall off. As the rotor's torque decreases, so does the current flow in the stator winding. When a force is applied to the rotor's shaft that is greater than the torque developed by the rotor, the rotor will begin to become more "out of speed" (it slows down) with the stator. As this occurs the amount of torque produced by the rotor and the current drawn by the stator windings will begin to increase. This is a result of the changes in the motor's impedance. That is, the torque required by the driven load and the current drawn by the motor are related over a given range. As the torque demand rises and falls, so does the current drawn by the motor.

Stator Construction

The stator is built up using thin steel sheets, called laminations, with slots in them. They are tightly stacked one atop another and then fastened together so that the individual notches form a continuous lengthwise slot on the face of the inside of the stack. Insulation is placed into each of these slots, then coils of a special type of insulated magnetic wire are wound with many coils and are inserted into the individual slots and connected together to form an electrical circuit. The completed stator assembly is pressed into the motor's housing and welded in place.

Two coils that are installed 180 degrees apart from each other form one pair of poles of the motor. Since a magnet has both a north and a south pole, the minimum number of coils a motor can have is four. That means that there will be four coils of wire in a two-pole motor. Motors are typically built with either two, four, six, or eight poles.

Rotor Construction

A rotor is similar to a stator in that it also has laminations that are stacked one on top of another. In place of copper wires, individual aluminum bars are installed in the rotor's slots. These conducting bars are connected together on each end by an aluminum ring called a *shorting ring*.

A ventilation fan is mounted on the end of the shaft opposite the load-driving end. The rotor shaft is supported in the center of the stator's housing by bearings on each end and an end bell on each side of the stator's housing. The inside diameter of the stator and the outside diameter of the rotor are slightly different to allow for a small air-filled gap. The rotor's bearings position the rotor in the center of the stator's magnetic field.

Figures 7–1 and 7–2 each show three sets of coils spaced 120 degrees apart around the stator housing. Since each phase is 120 electrical degrees out of phase with one another, the necessary change in current flow will occur in the same order, and in the same time gap between each other.

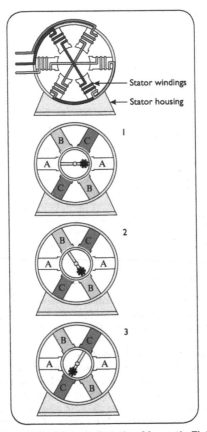

Figure 7-1 Motor's Rotating Magnetic Field

The figures show the sequence of changes in the magnetic field from starting to rated speed. The motor is started by applying power to all three phases. The coils marked A will produce a north and a south pole. The arrow and small black dot in the center of the graphic will move around

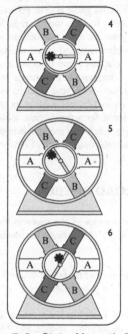

Figure 7-2 Stator Magnetic Fields

in a circle as the magnetic forces cause the rotor to begin to rotate. At 120 electrical degrees, phase B coils produce both a north and a south magnetic pole. The induced current in the rotor interacts with the stator's current causing the rotor to turn.

Next, phase C pole coils will produce a north and a south magnetic pole. The rotor is now following the stator's rotating magnetic field.

Next the A phase pole's magnetic field will again begin to build up, but this time the polarity of the two poles will be opposite, that is, the sine wave has crossed the zero line and is increasing in the negative potential direction. This change in polarity causes the magnetic field to continue to rotate, and the rotor continues to follow this field.

The next action will be for phase B to be powered up, magnetizing its pole in the opposite polarity to when they were in frame number 2.

Next, the C phase poles are again magnetized and the magnetic field has completed one complete turn, and so has the rotor. The stator's magnetic field changes because the current from the source has changed. The rotor current changes in response to the change in the stator's changed magnetic field. The interaction of these changes in polarity is what produces the force necessary to rotate the rotor at its rated speed.

A common rated speed for a three-phase motor is 1,750 RPM. To build a motor that will operate at this speed, it is necessary to install in the stator assembly, twelve coils, four in each of the three phases.

RATED FULL LOAD SPEED

The theoretical synchronous speed of a motor is based upon the rotating speed of its magnetic field. The formula for obtaining the synchronous speed of a motor is:

Speed = 120 × the frequency of the power source in Hertz ÷ the number of poles in the motor

Written in math terms, $S = 120 \times F \div P$, where S is speed in rpms, F is the frequency in Hertz of the power source, and P is the number of poles in the stator winding

Theoretical versus Actual Speed

When an induction motor is operating at its rated full load conditions, it will not be rotating at its theoretical synchronous speed. The rotor's speed and the stator's magnetic field's speed will differ by a small percentage, called *slip*. While the amount of slip differs from one motor design to another, it is generally between 1 to 5 percent of synchronous speed. The following lists theoretical and the average slip speed for one motor manufacturer.

Poles	Synchronous Speed	Actual Speed
2	3,600	3,450
4	1,800	1,750
6	1,200	1,140

SERVICE FACTOR

A service factor is normally listed on a motor's nameplate and is a multiple that can be applied to a motor's rated horsepower. A service factor of 1 indicates that the motor may not be overloaded without a reduction of service life. A service factor of 1.15 indicates that the motor can be subjected to a load 15 percent greater than its rated load without reduction of its service life. A service factor of 1.25 indicates that a 20-HP motor can be required to produce 25 HP without damage. A service factor should not be relied upon to continuously carry an excessive load. When conditions of voltage, frequency, and ambient temperature are not at rated conditions, the service factor is already being used.

RATED AMBIENT TEMPERATURE

Motors are commonly rated for 104 °F (40 °C). When a motor is operated 10 degrees hotter than its rated temperature, studies have shown that the motor's life will be reduced by 50 percent. Motors must be installed and maintained so that they can dissipate heat from its internal components. The outside surface of a motor is not the hottest spot of a motor. Not all of the heat generated internally is transferred to the outer surface of a motor. Some heat is transferred to the cooling air passing through ventilation openings in the motor. These openings and the connecting internal passages within the motor must remain open, or the motor's life will be reduced.

STATOR WINDING INSULATION CLASSES

The individual materials and methods of assembly used in the construction of a motor's stator windings work as a system that provides a specific level of electrical strength called *dielectric strength*. The four classes of insulation systems are listed below along with their maximum hot-spot temperature.

- ▪ Class A: 221 °F (105 °C)
- ▪ Class B: 266 °F (130 °C)
- ▪ Class F: 311 °F (155 °C)
- ▪ Class H is limited to 356 °F (180 degrees °C)

The maximum hot spot is typically located at some location deep in the stator's windings. When a motor is operated at temperatures below the insulation system's maximum rated temperature, the typical 20,000-hour rated service life will be increased. More importantly, excessive operating heat shortens a motor's life perhaps by as much as 50 percent over its rated maximum temperature.

TYPES OF TORQUE PRODUCED BY A MOTOR

The twisting force that a motor's shaft produces is called torque. The amount of torque produced when the rotor is not turning is called locked rotor torque. Each horsepower and type of motor will produce a different amount of locked rotor torque. The amount of torque produced by a motor as it increases speed is called pull-up torque. The maximum amount of torque that a motor will produce while running without a sudden decrease in speed is known as breakdown torque. The amount of torque produced by a motor at rated conditions is called rated torque, or rated load torque. Figure 7–3 is a plot of the torque developed by a motor over its operating speed. This is a general graph and is not typical of all three-phase motors.

By the study of Figure 7–3 it can be determined that full load torque is the lowest of all four torque values. For troubleshooting purposes, it should be understood that:

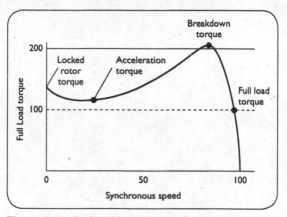

Figure 7–3 Typical Motor Torque Curve

1. If a motor cannot begin to rotate the load, the motor is overloaded.
2. If the motor cannot bring the driven load up to full speed, it is over-loaded.
3. If a motor is operating at rated speed, and then begins to slow down in speed, the motor is grossly overloaded.

8

CONTROLLING AND PROTECTING THREE-PHASE MOTORS

To control and protect three-phase motors a magnetic motor starter is typically used. An electromagnet (a coil of very small wire) is used to close the starter's three main power contacts. Starters can start-stop, speed up, slow down, change direction of rotation, and protect a motor from running overloads. Increasingly starters are provided with solid-state overload relays that provide the following types of protection:

- Phase loss
- Low voltage
- Phase reversal
- Voltage imbalance

COMPONENTS OF A MOTOR STARTER

Motor starters are composed of two major components: a three-pole (three switches that open and close at the same time) contactor which starts and stops the flow of electricity, and an overload relay, which senses current flowing to the motor. Overload relays have three sensors (heater elements), that sense power flowing in each of the three phases of the motor.

The main power circuit will pass power through the contactor, through the overloads, and to the motor, while the pilot control circuit directs power flow to the starter's electromagnetic coil that causes the main power to flow to the motor. While some motor starters use the same control circuit voltage as the power circuit, most do not. The most common control circuit voltage is 120 VAC. This voltage is supplied from a power source different from the one that supplies power to the motor.

MOTOR STARTER SIZES

The National Electrical Manufacturers Association (NEMA) has developed standards (rules) to ensure safety and reliability of electrical equipment for many years when properly applied, installed, and maintained. Table 8–1 provides a NEMA recommendation for the sizes of motor starters and the motors they provide power to.

Table 8–1 Motor Starter Size, Horsepower, and Voltage Chart

NEMA SIZE	Continuous Amp Rating	Horsepower at 230 VAC	Horsepower at 460 VAC
00	9	1	2
0	18	3	5
1	27	7	10
2	45	15	25
3	90	30	50
4	135	50	100
5	270	100	200
6	540	200	400
7	810	300	600
8	1215	450	900
9	2250	800	1600

The factors that determine the size of a motor starter are:

■ Single or three phase
■ Motor horsepower
■ Voltage

Size 00 is the smallest size starter and can supply power to either a 1-HP, 230 V motor, or a 2-HP motor operating on 460 V. A size 9, the largest, can provide power to a 1,600-HP motor operating on 460 VAC.

METHODS OF STARTING THREE-PHASE INDUCTION MOTORS

The most common method of starting a three-phase motor is with full voltage, across the line starting method. We will discuss this method in this chapter.

FULL VOLTAGE, ACROSS THE LINE MOTOR STARTERS

When a motor is started with the full voltage of the supply, it will draw a large amount of current (between five and eight times full load current). When a motor is running at rated speed and is fully loaded, it will draw full load (FLA), or rated load amperage (RLA). For example, a motor loaded to its rated horsepower pulls 10 A. When this motor starts across the line (XL), it will draw between 50 and 80 A depending upon the design of the motor for less than 10 seconds. When the motor reaches operating speed, the current draw will fall off to the motor's rated load amperage of 10 A. When this motor is pulling between 50 and 80 A there is nothing wrong with the motor; it is operating as it was designed to. However, the entire electrical system must be able to provide that 50 to 80 A for this brief period. The need to provide a large amount of power for a very brief time motivated the development of other methods of soft starting three-phase motors.

HOW A MOTOR STARTER OPERATES

Figure 8–1 provides a diagram of a control circuit for a typical motor starter showing how power comes from the source, listed as L-1, L-2, and L-3. These three wires connect to the main power connections marked on L-1,

L-2, and L-3. When the contacts are open, the power cannot pass to the motor. Figure 8–2 shows only the control circuit portion of the diagram shown in Figure 8–1. Notice that in Figure 8–2 there is a line, representing a wire, connected to the L-1 wire. This wire connects to one side of the SW-1 switch, which is open, so power is not able to flow to the motor starter's magnetic coil (M). Because this coil is not receiving power, it cannot produce a magnetic field. As there is no magnetic field, the main power contacts are held in their open position (off) by a small spring. Notice to the right of the magnetic coil, marked M, there is a symbol (OL). This symbol indicates that there is a normally closed switch in the circuit. This symbol is marked OL for overload protection. Notice that there is a wire connected to the right-hand side of the OL switch which is connected to L-2. Notice also to the right of each of the Ms, the main power contacts, there is a symbol that looks a bit like two fish hooks connected together. These are the symbols used for the current sensors (heater elements). These sensors have main power flowing through them. When the current flow is above the trip point setting, they will cause the OL switch to open the current flow to the magnetic coil. In summary, for the motor starter to close and the motor to run, the following must happen. There must be a voltage at L-1, L-2, and L-3. Switch SW-1 must close and pass power to the M coil. Power must

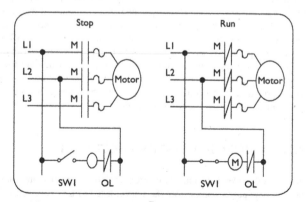

Figure 8–1 Motor Starter Schematic Diagram

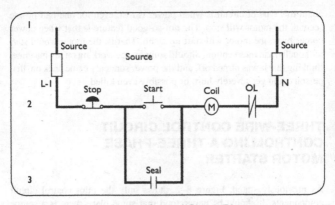

Figure 8–2 Motor Starter Control Circuit Diagram

flow from the M coil to the OL switch. Power must flow out of the OL switch and back to the L-2 connection point. Stated differently, power must flow from L-1 to and through the pilot controller SW-1, to the load, the M coil, out of the M coil, to the OL switch, out of it, and on to the L-2 connection. That is, power comes from the source, to and through the control, to the load, through the overcurrent protection (OL) and back to the supply source. The magnetic field overpowers the springs' closing the main power contacts passing power to the motor.

Notice the round circle with the word "Motor" written in it in Figure 8–1. This is the symbol used to indicate a three-phase motor. That is to say that everything works when there is power flowing, or a voltage and a current flow in the circuit. When the SW-1 switch is open, power cannot flow to the M coil, so it cannot create a magnetic field, so the main power contacts remain open. When there is no magnetic field, the springs in the starter keep the main power contacts open, so power cannot flow from the supply source to the controller, the M contacts, through the overload sensors, and on to the motor.

Notice that the control circuit has two wires, one on each side of the pilot control—the SW-1 switch. When there are two wires in the control circuit, it is commonly called a *two-wire control circuit*. A good point about this type of circuit is when power is lost, even for one-half of one second, the motor will stop. The not-so-good feature is that when power comes back, the motor will start up again. That is, the motor would start without human intervention. Should someone be working on the machine, thinking that it was turned off, and the power suddenly came back on, that person could get severely hurt, or possibly even killed.

THREE-WIRE CONTROL CIRCUIT CONTROLLING A THREE-PHASE MOTOR STARTER

As previously stated, Figure 8–2 shows only the pilot control circuit components. It should be understood that some place there is a source of power, and it is marked L-1. Power flows from the L-1 source to a protective device, in this case a fuse (the rectangular box), then on to the stop push button (pilot control). The stop push button has a spring in it that maintains the switch contact in the closed position. So with the stop button closed (its normal state), power is passed to one side of the start push button (another pilot control). The start push button also has a spring in it but its spring holds it in the open position—the non-power-passing position. Pushing the start button will pass power to the motor starter's magnetic coil, and then to the contacts marked OL and on back to the supply source. The coil will create a magnetic field and the starter's main power contacts will close, passing power to the overload sensors, and on to the motor, and the motor will run.

Now if this contact is closed, power will flow back to the power supply source (L-2). In order for the motor to continue to run, someone must keep pushing on the start push button, for as soon as it is released, its internal spring will open the contacts, and control power will stop flowing, and the motor starter's contacts will open, and the motor will stop running. Notice that there is a wire connected between the stop and start push buttons, and that it connects to one side of a switch marked "seal." This is a normally open (N.O.) set of contacts physically linked to the main power contacts of the motor starter. That is, when the M contacts pull in, this seal switch closes.

When the start push button is pushed, power is passed to the coil, and the M contacts close, and the seal contacts also close. When the start button is released, the seal contacts are still closed, so power is not interrupted; it still passes from one side of the stop button, to and through the seal contacts and back to the other side of the start button. The result is the motor will continue to run when the start button is released.

There are three wires in this control circuit. One going to the stop button, one between the stop and start push button, and one on the other side of the start button; this is why the use of a stop-start button arrangement is known as a three-wire control circuit. The contact marked "seal" provides "a seal in" circuit that serves as a path for power to flow when the start button is released.

What would happen if power were lost for, say, one-half of one second? No power would be flowing to the seal in contact, so power would not flow to the M coil, and the starter would "drop out." That is to say, it would open its M contacts, and the motor would stop running.

To restart, a human would have to intervene, that is, push the start button. This is a safer circuit. A three-wire control circuit is also known as a *no voltage release* circuit. When the voltage is lost, even for a fraction of a second, the motor will stop running.

SIZING OF OVERLOAD HEATER ELEMENTS

Overload heater elements are rated by trip current. The National Electrical Code directs that either 115 or 125 percent of the motor's nameplate FLA be used for sizing overload heater elements.

- When the SF listed on the motor's nameplate is 1, use the 115 percent value.
- If the nameplate is marked temperature rise of 104 °F (40 °C), or less, use the 125 percent value.
- If the SF is marked 1.15, use the 125 percent value.
- For all other motors use the 115 percent value.

Stated another way, use the 125 percent value for motors marked with an SF of 1.15, or when the motor is marked with a temperature rise of 104 °F (40 °C), or less. Use 115 percent when the SF is marked 1. If the motor is not marked "Duty: Cont.," use the 115 percent value.

By multiplying the FLA or RLA value marked on the motor's nameplate by either the 115 or the 125 percent, the heater element size will be determined. The heater tables are already increased, do not oversize the heaters by adding 115 or 125% again! Using the starter manufacturer's heater table (typically located inside of the starter enclosure), select the heater element number that matches the value above.

By following this procedure, the motor will trip-out whenever the running load amperage exceeds the nameplate running load amperage by either 15 or 25 percent. When a new NEMA-rated starter is purchased, the heater elements are not included with the starter. They must be selected and ordered separately. The elements cannot be selected based upon the NEMA size of the starter, or the motor horsepower. The heater elements must be selected by following the procedure listed above using only the FLA/RLA amperage marked on the motor's nameplate.

VARIABLE FREQUENCY DRIVES

A variable frequency drive (VFDs) uses solid state electronic circuits to provide a signal to a motor for the purpose of controlling its rate of acceleration, deceleration, direction of rotation, torque, and horsepower. Drives do this by converting a fixed frequency (either 50 or 60 Hz) AC power source, seen in Figure 9–1, to a variable frequency power supply to the motor, as seen in Figure 9–2.

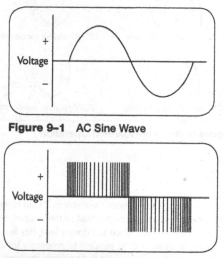

Figure 9–1 AC Sine Wave

Figure 9–2 High Frequency Signal

Drives have proven to reduce peak energy demand, decrease maintenance cost, and improve operating reliability.

FOUR-QUADRANT OPERATION

A drive can be required to speed up or slow down a motor and change the rate at which the motor is adjusted. A drive can also be required to change the direction of rotation of the motor. Restated, a drive can be required to operate the motor in a counterclockwise (CCW) direction or in a clockwise (CW) direction. That is, a drive can be capable of operating in one or more of four quadrants: CW, accelerating or decelerating, or CCW, accelerating or decelerating. Figure 9–3 lists these four quadrants.

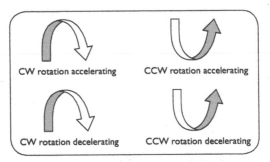

Figure 9–3 Four-Quadrant Operation

TORQUE

Motors must develop torque to move the connected load from a dead stop (breakaway torque) and accelerate the load to the desired rotating speed (acceleration/pull-up torque). Once the driven load has reached normal operating speed, the motor will be required to produce a lesser amount of torque (rated torque) than was needed to accelerate the load. Some loads will demand that the motor produce a short-term rapid increase in torque to overcome a higher demand by the load. This occurs with machines such as hammer mills. Machines are grouped into two general classes, constant and variable torque loads. Centrifugal pumps and blowers are examples of variable torque loads. Cranes and coil winders are examples of constant torque loads.

Three-phase induction motors are capable of developing an amount of torque greater than rated load torque. Breakaway torque is a greater than rated torque; rated torque is less than breakdown torque. When selecting a replacement drive it should be confirmed that the drive is capable of supplying the torque demanded by the driven load across its entire operating range and not a single point.

Speed Regulation

When a transient load is applied, a motor will slow down between 3 and 5 percent. Modern drives can hold the speed of a motor to within 1 percent of set point. Many vector control drives can hold the speed of the motor to within 1/10 of 1 percent of the set point.

Torque Development

When a motor starts up it must develop enough torque to get the load moving. This is called breakaway torque. In order for the motor to produce this high level of torque, it must increase the force of the magnetic field. To do this, it increases the current drawn by the motor. When a motor starts at full rated voltage, it draws between five and eight times as much power as it does when it is running. Start-up current is called inrush current. All of the electrical system components in the circuit must be sized to handle this increased amount of power.

A VFD can reduce the initial start-up inrush current. A drive does this by reducing the voltage to the motor when it first begins to accelerate the load. This lower voltage reduces the motor's current draw. As the motor starts to rotate the load, the drive senses this movement and increases the voltage, allowing the amperage to rise, increasing the strength of the magnetic fields. The drive continues to sense the rotor's increase in speed and in turn increases the voltage supplied to the motor's windings. The drive also changes the frequency of the applied voltage, allowing the amperage to rise, increasing the strength of the magnetic fields. The drive continues to sense the rotor's increase in speed and it increases the voltage supplied to the motor's windings.

The drive also changes the frequency of the applied voltage as the motor increases its speed. The drive continues to increase the voltage and the frequency until the motor is operating at the preset speed. The drive can also raise the voltage applied to the motor above its rated voltage to allow additional torque to be developed during start up and under heavy loads. The relationship between the voltage supplied to the motor and the frequency of the electrical cycle is known as the *voltage to Hertz ratio* (V/Hz). That is, the voltage supplies the torque needed and the frequency controls the speed at which the motor is operated.

Control of Torque and Speed

When the load's torque demand decreases, the motor will want to increase its speed. To maintain a constant speed, the drive will decrease the frequency of the power applied to the motor. Should the load increase, the motor will want to slow down. The drive can increase the frequency to speed up the motor. To keep torque developed by the motor at a constant, after the driven load's torque has increased, the drive can decrease the frequency—this will keep slip constant thus maintaining the motor's torque output. Should the load torque demand decrease, the drive can increase the frequency of the electrical pulse supplied to the motor. Thus it can be said that modern VFDs respond very well to the requirements of the load.

PARTS OF A DRIVE

Enclosure: The external housing is the part that protects people from accidental contact with energized parts. It also protects the drive's electronic components from environmental contamination.

Terminal board: Two are provided, one for line and one for load conducts. Additional terminal boards are provided for connection of the various input and output signals.

Key pad: Push buttons or key pads are used for entering and obtaining data from the drive.

LED: Drives will normally have LEDs providing visual communication with the operator.

COM: The drive may have the ability to communicate with a remote process control network, or an intranet and Internet.

External inputs: The drive will have terminals for connection to external devices such as remote start-stop stations, set point, reverse, jog, and alarms.

Inverter-converter: This section changes the AC input power to DC and then back to a variable AC signal. This is many times accomplished with rectifiers or diodes. Rectifiers and diodes allow power to pass in one direction only. This results in an output signal that looks more like a ripple than that of a pure AC sine wave.

DC Bus Section

The DC bus in a VFD smoothes out the ripples in the pulsed power produced by the converter section. Capacitors and other electronic devices are combined to smooth out the ripples in the pulse power produced by the converter section. Some drives require a variable voltage to the DC bus. In these drives, the rectifier bridge may chop the pulses into smaller pieces using a chopper circuit. The larger the pulse produced by the circuit, the higher the voltage. The smaller the chop, the lower the voltage supplied to the DC bus.

Inverter Section

The inverter section switches the DC signal on and off. This on and off switching is used to piece together what looks like to the motor an AC signal. This is accomplished by using either silicone controlled rectifiers (SCR) or power transistors. Insulated gate bi-polar transistors (IGBTs) that are used in many drives today can switch at a rate of about five million times per second. This allows the drive to produce a waveform that is much closer to a standard AC signal.

PULSE WIDTH MODULATION

The majority of drives manufactured today use a method called *pulse width modulation* (PWM) to invert the DC pulse back into an AC sine wave. On three-phase drives six IGBTs are used—three for the positive bus and three on the negative DC bus. With PWM the voltage is pulsed on and off in sequence to produce a three-phase signal. The pulses are wider as the voltage is increased, and narrower as the voltage decreases. This allows the drive to control both voltage and frequency signals.

SIX-STEP METHOD OF DRIVE CONTROL

The six-step method uses SCRs in the inverter section to switch the voltage from positive to negative in an offset pattern of pulses in each of the three phases. When this method is applied to the motor, the pulses create a pattern that simulates the rising and falling of the current of a phase of an AC sine wave. The drive changes the length of time the SCRs are on. This changes the frequency of the signal to the motor. By shortening the on time, the frequency to the motor is increased. When the drive lengthens the on time, it decreases the frequency of the signal to the motor. Voltage to the motor is changed in the converter. This is accomplished with either a rectifier or chopper circuit.

Soft Starters as Torque Controllers

When a motor starts it draws locked rotor amps (LRA). This can be between five and eight times the motor's running full load amps. This large inrush current draw is necessary for the motor to produce sufficient breakaway torque to overcome the inertia of the rotor and the driven load. Motors that are started with full voltage reach rated speed quickly. This quick acceleration jolts both the rotor and the driven load.

The serving electric utility must provide enough power to supply these peak power demands. Power companies bill commercial customers a demand charge for these peak power demands. Electromechanical methods of starting motors use several different methods of reducing the inrush current draw of the motor. Some use resistors, others transformers. Solid state drives are more reliable than older types of reduced voltage starters. Single speed motors can be used in place of more expensive multispeed starters. Electromechanical starters still give a jolt to the drive train. The opening of contactors during starting can create voltage spikes in the electrical distribution system within a facility and will likely have a negative impact on sensitive electrical equipment.

PROGRAMMING DRIVES

Drives can be programmed by the operator. This allows the operator to preset how the drive operates. Many drives allowed this to be done through a menu-driven keypad operator interface program. These programs differ from one drive manufacturer to another and between the same manufacturer and individual models. These keypads are sometimes numbers only and sometimes alpha numeric. On some brands there is a local keypad and display, while on others a desktop or handheld PC can be used to program the drive. Still others use an EPROM, or EEPROM.

Drive Presets

Many drives today have features that allow it to be preprogrammed to operate. The menu commonly offers several options from which to select. Some allow a value within a given range, while others allow only one of a few values to be selected. The following is a list of the features that can be set by the operator on several brands of drives available today.

Frequency

Different frequencies can be selected so that the motor can be operated at different speeds during each step of the process cycle. There may be both high and low frequency limit set points. A word of caution: many motors will not operate at extremely slow speeds without overheating, so be careful that the low speed setting is high enough to allow the motor to operate within its designed limits. Commonly motors will run too hot below 10 Hz, while some will operate as low as 5 Hz.

Frequency Avoidance

Machines that rotate have what is called a critical speed. Most everything mechanical has a natural frequency. When rotated at its lowest critical speed, the machine will resonate, which results in a significant increase in vibration. A drive should not be allowed to operate for prolonged periods at the machines' resonate frequency. For this reason, some drives allow the operator to select a specific frequency at which the drive cannot operate. Most squirrel case blower wheels have two critical speeds, the first can be passed through as the load accelerates-decelerates. The second critical speed will result in the blower wheel flying apart. The drive should not be allowed to operate the blower wheel at either of these speeds.

Jogging Frequency

When a motor must be operated in a jogging mode, the drive may have a selectable frequency for this task. Typically, this is between 5 and 10 Hz.

Acceleration-Deceleration Times

Sometimes this is called ramp-up and ramp-down rates. These are rate of speed changes the motor will follow. This is a key element in reducing the amount of sudden jolt delivered to a drive train by the motor. Many drives offer options of more than one set of acceleration-deceleration ramps of rates.

Additional Features

The features available from drive manufacturers vary. The following is a short list of some common ones.

- Jog frequency
- Acceleration-deceleration rates
- Braking currents
- Volts to Hertz (V/Hz) ratio
- Frequency set points

10

PROGRAMMABLE LOGIC CONTROLLERS

This chapter covers the basic abilities and styles of programmable logic controllers (PLCs). To smooth the transition from electromechanical relay–based systems to PLC-based control systems, PLC's were originally programmed using diagrams that looked like standard electrical ladder diagrams. Today, many PLCs use various high-level programming languages in addition to basic relay logic.

BLACK BOX CONCEPT

Consider a black box setting on a table. Into this box a cup of water is poured, along with two spoons of a brown powder. It is then connected to a 120-VAC outlet.

Figure 10–1 is a block diagram of a PLC. This black box has four small lamps colored red, green, yellow, and blue. When connected to the 120-V power source, the red light comes on.

When one cup of water is poured into the black box, the blue light comes on. When two spoons of the special brown powder are added, the yellow light comes on. Three minutes later the green light comes on and coffee fills a small cup. The black box machine has done what it was designed to do. It has made a cup of coffee. It is a coffee maker. To make coffee it needs inputs, which are water, coffee powder, and 120 VAC energy.

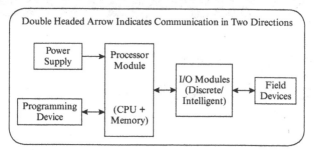

Figure 10–1 Input-Processor-Output Diagram of a PLC

The product that it makes (its output) is one 6-ounce cup of coffee. The black box needs specific inputs to provide an output. How the black box works inside is a mystery. However, how to use it to make coffee is simple—input water, coffee, 120 VAC, and in three minutes out comes coffee. When the input 120-VAC is present, the red light comes on. When water is poured into the machine, the blue light comes on. When coffee is added, the yellow light comes on.

When the coffee is brewed, the green light comes on. It is easy to tell when the input water has not been added—the blue and green lights will be off. No water, no coffee. When the input coffee is absent, the brown light will be off.

To troubleshoot the black box, look to see if it has the required inputs. When the proper inputs are provided, the designed output will be produced after a three-minute process delay.

The logic of the machine is simple, inputs results in outputs. If it does not work, check to see if it has the needed inputs.

The black box (PLC) can be programmed to complete a specific task in a specific order, in a specific amount of time; given specific inputs, it will produce specific outputs. With PLCs inputs and outputs are low-power electrical signals, not water and coffee.

In brief this is how PLCs work. They are logical solid state electrical machines, and given the required inputs and internal instructions they will provide visual indications (lights) that the needed inputs are present.

When the internal program has turned on the electric water heater, and the timer has timed three minutes, the program will turn on the green light (a visual output) and coffee will fill the cup. Figure 10–2 shows a few of the hard wired real world inputs and outputs typically used with PLCs.

The lights are outputs, they provide visual indications that the needed inputs are present, and that the desired output, a cup of coffee, is available in the cup. Once the cup is filled the machine resets to begin the cycle again.

PLCs have the following major components:

▓ Electrical power for the microprocessor, heater, and inputs/outputs
▓ A place to store the instruction program (memory)
▓ A section to connect the electrical inputs connections
▓ A section to connect the electrical outputs connections
▓ Housing to protect the components
▓ Some means of programming the PLC

PLCs are dumb in that they know only the information they have received (inputs) and respond only in the manner they have been instructed to (programmed to).

The processor (the brains) scans (reads) inputs, and stores the condition (on or off) of the inputs in memory. Then the processor executes its instructions or program. It goes to the memory and obtains the current condition of all of the inputs, and places their state in the proper location in the program,

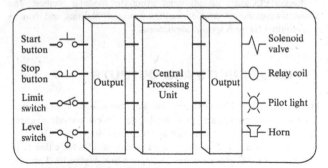

Figure10–2 PLC Components

and then makes decisions, and changes the status of the various outputs. The program operates in a continuous cycle—it reads inputs, executes the program, and changes specific outputs, over and over until it loses power, one or more vital inputs, or an error occurs in the program.

PLC SIZES

Small PLCs will typically have up to 32 inputs and outputs (I/Os). Medium-sized PLCs have between 32 and 128 I/Os. Larger PLCs may have as many as 8,000 inputs and outputs. Generally as the number of I/Os increases, so does the programmability of the PLC increase. Virtually any electromechanical device can be obtained as a software or logical equivalent. On and off delay timers and counters are routinely available as a logical equivalent.

PLC CONSTRUCTION

PLCs come in two general styles of construction brick and modular. Smaller PLCs are made as bricks with all of the components located within a single housing and mounted on a single printed circuit board.

Larger PLCs are manufactured using the modular method. This allows for specific selection of the I/Os, processor abilities, and memory size required for each specific application.

TYPES OF INPUTS AND OUTPUTS

I/Os are classified as being discrete (digital), that is, capable of providing a signal, or not providing a signal. Analog devices can provide a variable signal across its limited range such as 4 to 20 milliamps (mA) DC or 0 to 10 VDC. Typically, DC digital voltages used with I/Os are either 5, 12, 24, 48, or 60 V. AC I/Os are either 12, 24, or 120 V. Figure 10–3 provides several of the symbols used in programming PLCs.

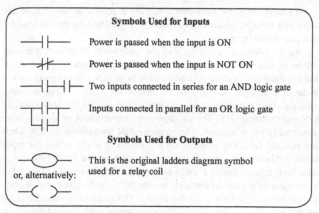

Figure 10–3 PLC Inputs and Outputs

Modular I/Os

Considering PLCs that utilize the modular method of construction, a given number (4, 8, or more) inputs or outputs are contained on an individual slide in module. All of the I/Os on a module will normally be of the same type, either analog or digital, and of the same voltage.

Contacts as Inputs

The simple closing of a set of switch contacts that pass a voltage to the terminals of an input on an input module represents a change in a measured variable, such as the closing of a hopper's fill bin door. While there are many types of advanced PLC programs in use today, most use a modified version of a standard electrical ladder diagram. On such a diagram a symbol is used to represent the real-world switch contacts, as being either normally open or normally closed. Figure 10–4 provides diagrams of the On-Off states of PLC real world (as opposed to the logic devices used by the PLCs program) inputs. A change in the position of the specific input causes a change in the status of that bit of information that stands for the real-world condition.

A change in some measured condition is passed to the PLC in the form of a voltage, which in turn is sent to a specific memory address for that real-world input.

As the program scans each of the addresses, it updates the condition of that bit of information in the PLCs program. When all of the required preconditions (inputs) have been met, the program makes a decision to take action. That is, change the status of one or more outputs. Two common types of instructions for PLCs are: examine ON and examine OFF. The examine ON instruction is associated with a normally open contact. The examine ON instruction is ON when the input is ON. The examine OFF instruction is ON when the input is OFF. The various logical instructions such as the AND instruction (the logic equivalent of a series circuit), the OR instruction (the logic equivalent of a parallel circuit), and the NOT instruction, means that the output will be OFF when the input is ON can be organized into all of the possible combinations. When combined with logical functions such as timers and counters, any complex process can be controlled with a PLC.

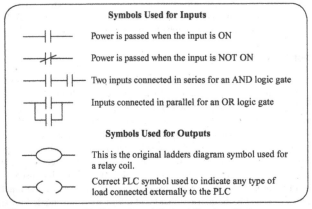

Figure 10-4 PLC Input and Output Device States

Coils As Outputs

Most PLC ladder diagrams use a modified version of the symbol used in the original (pre-PLC) electrical ladder diagrams, that is, a circle with two short horizontal lines. PLCs typically use a circle that is about one-fourth open at the top and bottom.

This symbol indicates some type of load—an electrical energy consumer—that does useful work. That could be a relay coil, a solenoid coil, an electric motor, and so on.

The coil symbol on a PLC diagram is a symbol for some type of load and does not always indicate the coil of a relay.

The PLC's processors scan addresses, update conditions, compare the current "is" condition with the "should be" condition and then change the status of one or more outputs. PLCs do a lot of comparing of the "is" condition to the desired condition (program). Modern PLCs can do much more than compare one condition to another.

TIMERS

Timers are used to extend, compress, or shift the occurrence of an event. To provide this control function, PLCs are routinely provided with one or more logic timers. The three most common timers are on-delay, noted on the program by the use of the letters (TON); off-delay, noted by use of the letters (TOF); and retentive timers, noted by the use of the letters (RTO).

PLC programs provide the timer's address, the time base, and the timer's preset value. The address is the location in memory where the timer's input signal is sent. The time base of a timer determines how fast a timer will count. For example, a time base of one second instructs the timer to count individual seconds.

A value called a preset must be entered into the program. Say the number 500 was entered as the preset. When the input closed, the timer would count in seconds until 500 seconds had passed, and then the contacts would change state. That is, the program rung would become true and its output would be energized.

Viewing the program the amount of time accumulated will be displayed on the program diagram in time base units that have elapsed. When speaking of electromechanical timers, it will be said that a timer has "timed out." When speaking of PLC timers, it will be said that a timer is "done" (noted as DN on the program). To clear the accumulated value it is necessary for the reset instruction to be true. Then the value in that memory location will be rest to zero. The two most common timers used in PLCs are On-delay, and Off-delay. With the On-delay timer, the output is on for some time period after the input is on, and turns off immediately when the input is turned off.

COUNTERS

For some processes the occurrence of something would need to be counted, perhaps a preset number of bolts. Say the process is filling a box and then closing the lid.

A counter would receive a signal every time one bolt passed through the field of view of a noncontact proximity sensor. The two most common types of counters are count up (noted on the program as CU) and count-down counters (noted as CD on the program). When the counter has counted up to the preset desired value, the letters DN will be seen on the program rung for the counter.

A counter looks at changes in the input signal. For the bolt counting example, as each bolt passes through the field of view, the signal changes state, and the counter increments in accumulated value. When the count reaches the counter's preset value, the program rung will be true and the output assigned to the counter will be energized.

Like the timer, the accumulated value must be cleared, that is, reset. This is accomplished by making the remaining preset conditions become true.

BENEFITS OF USING PLCS

When changes are required in how a non-PLC-controlled process operates, it may be necessary to remove some existing electromechanical devices and possibly add others and to remove and install individual wires. In order to do this it is necessary that the process be stopped and placed in an electrically safe work condition. Then the rewiring can safely begin. Components can be removed, added, and generally rewired as and where needed. All of which must be physically done on the plant floor, under the same environmental conditions as the machine.

With a PLC-controlled machine, a designer can sit in an air conditioned office without danger of getting shocked and develop the program using either a hand-held programmer or a desktop computer. When completed, the new PLC program can undergo a simulated test run on a desktop computer without fear of damaging raw materials or exposing plant personnel to unexpected and unplanned machine behavior.

When the machine is controlled by only electromechanical components, there is a possibility that mistakes will be made in the wiring of the actual components.

With the PLC program, a designer can have a high degree of confidence that the new program will operate as designed, because it has already been tested and debugged as necessary. That is not to say that existing hardware on the plant floor will not need to be changed, added, or rewired.

I/Os AND MEMORY ADDRESSES

The status of inputs and outputs and the control program are all stored in the PLC's memory. The memory unit is typically either a battery backup or a nonvolatile type. "Nonvolatile" means that should power be removed the contents of the memory are not lost.

Memory size is typically listed in units such as 1 kilo, for 1,024 bytes; 1 M (for one megabyte), 1,048,576 bytes; or 1 G (for 1 gigabyte), 1,073,741,824 bytes.

Word size varies in multiples of 8 from 16 to 32 to 64 bits. A bit is a standard address for a file to be stored in memory.

A byte is the smallest unit of storage that can be assessed by a PLC's memory. Generally, PLC processor memory is divided into the following areas: I/O image tables, internal controls, registers, counters, timers, and the control program. The sizes of all of these consumers of memory vary depending upon their individual memory specifications.

11

ELECTRICAL SERVICES

An electrical service connects the facilities'/premises' wiring system to the serving electrical utilities' wiring. Figure 11–1 provides two diagrams: the upper one for the utility transformer connected in wye pattern, and the lower one for the point where the utility system supplies power to the service at the building. Figure 11–2 provides the various voltages common to three-phase 208/120 systems. The following are the exceptions to the general rule specified by the National Electrical Code (NEC)—that a single building may have only one service:

- Fire water pumps
- Emergency Systems
- Optional and legally required standby systems
- Parallel power production systems
- Systems designed for connection to multiple sources of power supply for the purpose of enhancing reliability
- A building of sufficient size as to warrant the use of more than one service
- Where special permission is obtained from the authority having jurisdiction

The conductors over which power flows can be installed from poles where they are called *overhead service entrance conductors;* when routed underground they are called a *service lateral.*

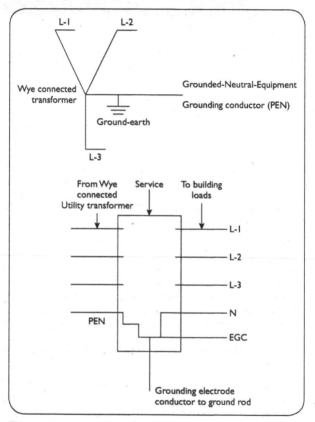

Figure 11–1 Solidly Grounded Three-Phase Wye-Connected Transformer and Service

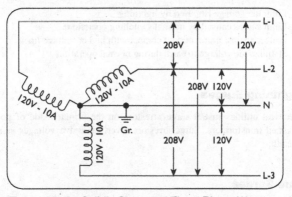

Figure 11–2 Solidly Grounded Three-Phase Wye-Connected Transformer Voltages

METERING EQUIPMENT

Where the load is less than 200 A, the billing meter has the full load current passing through the meter. Where the voltage is above 600 V, the output from current transformers (CT) is connected to the billing meter (kWH meter) which records the power consumed.

These small transformers produce an output voltage on their secondary that varies directly with the current flowing through their primaries, which are the service entrance conductors. To reduce operating cost, electric utilities are increasingly using power line carrier and wireless "smart meters," which use the power lines to send energy usage data to the utility.

GROUNDING OF ELECTRICAL SYSTEMS

Grounding an electrical system requires that an electrical connection be made from the system to earth. In Europe, grounding is called "earthing." Not all electrical power distribution systems are grounded. Grounding serves to:

- ▨ Limit the voltage imposed by lightning
- ▨ Limit surges caused by normal switching operations
- ▨ Protect against unintentional contact with higher-voltage lines
- ▨ Stabilize the voltage to earth during normal operation

Lightning Surges

Electrical utilities install surge arrestors on the primary side of pole-mounted transformers. Surge arrestors divert excessive voltages safely to earth.

Line Surges

The normal operation of a power distribution system requires that adjustments be made to generators, transformers, reactors, capacitors, and reclosers. All of these can cause variations in the system's operating voltage. When the system voltage becomes too high, the excess is routed to ground by virtue of the excessive electrical pressure (voltage) passing over the system grounding conductors.

Unintentional Contact with Higher-Voltage Lines

With overhead power distribution systems, the higher-voltage conductors are usually installed at the top of the poles. Lower-voltage conductors are installed below the higher-voltage conductors.

When poles are broken by impact by an out-of-control automobile, or when tree limbs fall on high-voltage conductors, they may come into momentary contact with the conductors of a lower-voltage system. This can result in a high voltage being imposed upon the lower-voltage system. By grounding the lower-voltage system, a low impedance (AC resistance) path is provided for the higher voltage to be safely dissipated in the earth.

Grounding has been proven to reduce the frequency and degree of damage but it is not a guarantee that the system will be unharmed.

Conductor Capacitance Leakage Current and Bonding

All electrical insulation materials leak some quantity of electrons. When a system is grounded, all conductive parts of the system and any connected appliances are bonded to the system grounding conductor, and in turn, to the grounding electrode conductor and to the grounding electrode (ground rod).

This creates a low impedance path for normal safe, low-level capacitance leakage current to be drained off before it can build up to the point of possibly creating a spark of sufficiently high voltage to result in someone being shocked or starting a fire.

BONDING VERSUS GROUNDING

Bonding is making an electrical connection from any normally non-current-carrying conductive surface to the system grounding point, the ground rod. In addition to the electrical supply, telephone systems, water piping, building steel, metal conduit, and metal enclosures are required by the NEC to be bonded to the electrical system's ground rod.

FAULT CURRENT PATH, CONDUITS, AND LOCKNUTS

Where the available short circuit current (SCC) is high, the conduit thread or its locknut can be burned open before the circuit breaker has opened to protect the circuit. This results in an uncleared ground fault that is dangerous to personnel. As a result, the NEC has rules for what types and sizes of conduits/raceways can and cannot be used as a fault current path.

Article 250 of the NEC requires that at "services" connections using reducing washers, oversized concentric or eccentric knockouts shall be supplemented with a bonding jumper.

LOW AND HIGH IMPEDANCE GROUNDED SYSTEMS

Electrical systems are classified as ungrounded, low impedance solidly grounded systems, and high impedance grounded systems.

When a short to ground occurs on a low impedance solidly grounded system, the impedance of the fault current path and the system current capacity are the factors that determine the amount of current that can flow. Depending upon the capacity of the system, the amount of short circuit fault current can be capable of producing major damage.

By inserting a resistor or reactor in the grounding electrode conductor where it connects to the grounding electrode, the amount of current that flows under short circuit conditions can be greatly reduced. This type of system is known as an impedance grounded system.

Systems with a phase-to-phase voltage of less than 1,000 VAC are normally solidly grounded. Medium voltage systems (1,000 to 15,000 V) can be "low" impedance grounded, limiting the fault current to about 25 A. High impedance grounded systems typically limit fault current to about 10 A. In industrial mining where portable substations are routinely used, use an impedance grounded electrical system.

UNGROUNDED ELECTRICAL SYSTEMS

The following systems are commonly ungrounded:

▪ 240-V, three-phase, three-wire, delta-connected
▪ 480-V, three-phase, three-wire, delta-connected
▪ 2,300-V, three-phase, three-wire, delta-connected
▪ 4,600-V, three-phase, three-wire, delta-connected
▪ 13,800-V, three-phase, three-wire, delta-connected

Figure 11–3 is a diagram of an ungrounded three-phase delta system without the voltages listed. Figure 11–4 is a diagram of a three-phase ungrounded delta with typical voltages listed.

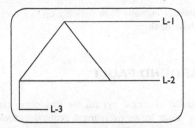

Figure 11–3 Ungrounded Delta-Connected Transformer

An ungrounded electrical system is provided with an equipment grounding conductor but not a current-carrying grounded conductor. With these systems, a connection to earth of all conductive non-current-carrying metal enclosures, motor housings, and metal conduits does not provide a fault current path. When the electrical system does not have a grounded conductor there is no fault current path to ground.

This results in the system's circuit protective devices not opening the circuit underground fault conditions. This is because the current cannot return to the system over the equipment grounding conductor. There is no connection between earth and the un-earthed electrical system. Bonding

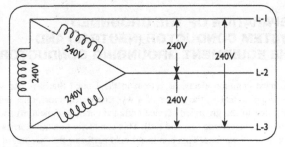

Figure 11–4 Ungrounded Delta-Connected Transformer Voltages

prevents small leakage currents from building up to a potentially dangerous voltage above earth.

SECOND GROUND FAULT

With an ungrounded electrical system the first phase that faults to ground does not result in an overcurrent protective device (OCPD) opening to protect the circuit. Should a second fault to ground develop, then the fault current can flow back to the source (transformer). The current level will be determined by the fault current path's impedance and the amount of fault current the system can develop. This should and normally does result in operation of the circuit's OCPD. That is, on an ungrounded system when the first ground fault develops, the OCPD will not operate as there is no path for the fault current to take back to the source of supply. If the first fault is not cleared, should a second fault to ground develop, there will be a path over which the fault current can flow back to the source.

Draining off low levels of insulation capacitance leakage current by grounding the normally non-current-carrying conductive surfaces of equipment only eliminates the build-up of static electricity on metal parts of the system during normal operation. It does not ground the system.

SEPARATION OF THE GROUNDED SYSTEM CONDUCTOR (NEUTRAL) AND THE EQUIPMENT GROUNDING CONDUCTOR

The most common electrical system in the United States is one with the midpoint, that is, the center of a wye-connected transformer that is connected to the grounded service entrance conductor (neutral) and to a grounding electrode (ground rod). This conductor (neutral) carries the imbalance of the load. This conductor also performs the function of the protective earth (PE), and is called the *protective earth neutral* (PEN). When this conductor reaches the service enclosure, it is connected to the neutral bus and to the ground bus along with the grounding electrode conductor and all equipment grounding conductors (EGCs).

From the service enclosure only the neutral conductors are connected to the neutral bus. The service enclosure is bonded (typically by a strap or screw) to the neutral bus.

The neutral and the EGC are always maintained separate from each other throughout the system. The neutral carries current during normal operation of 120-V circuits, and the load imbalance current for all multi-wire 240-V circuits, where three wires, line one, line two, and the natural, are provided. The EGC carries current only during ground fault conditions. The EGC is not to be used to carry current under nonground fault conditions. The EGC normally carries only minor capacitance leakage current. All circuits have some amount of leakage current. No insulation material is a perfect insulator. All leakage current is safely routed to the ground rod and is dissipated into the general mass of the earth.

WHAT CONDUCTORS MUST BE GROUNDED

With a single-phase two-wire system, one conductor must be grounded. This results in three conductors—a hot, a neutral, and an equipment grounding conductor being provided. For a three-wire system, the neutral conductor must be grounded. With a wye-connected system the midpoint or center tap must be grounded. Where the midpoint of one phase of a delta-connected system is grounded, the neutral becomes the grounded midpoint conductor.

For a conductor to be considered a neutral, there must be at least two conductors of opposite potential between each other present. Typically that will be 240 V, with one conductor that has a potential difference between itself and the two other conductors. In a typical 120-V outlet there is only one hot conductor present. Technically speaking the grounded conductor (white insulated wire) is not a neutral, as two hot conductors of opposite potential are not present.

Considering AC systems with a nominal operating voltage between 50 and 1,000 V, the circuit must be grounded if even one of the following conditions is true:

▧ The ungrounded conductor voltage is no more than 150 V.

▧ The system is a three-phase, four-wire wye-connected system with a neutral conductor

▧ The system is a three-phase, four-wire delta-connected system and the midpoint of one phase is used as a normal current-carrying conductor

▧ The grounded service entrance or underground service lateral is an uninsulated conductor, as is the case with most residential and light commercial facilities.

SEPARATELY DERIVED SYSTEMS

To understand the concept of a separately derived system (SDS) consider a transformer, composed of two physically and electrically separate, magnetically connected circuits. A transformer that is grounded on the primary is not automatically grounded on the secondary side. And a transformer that is grounded on the secondary side is not automatically grounded on the primary side. When all of the circuit conductors, including the neutral conductor, have no connection with the normal supply equipment such as generators or UPS systems (typically a battery back-up), the system is considered to be a separately derived system.

Quickly stated, if the neutrals of the normal source and alternate source (local generator) are hard wired together, (that is, there is no switch in the neutral path), the system is not a separately derived system.

STANDARD VOLTAGES

Single-phase services are commonly available at a nominal voltage of 120/240, three-wire grounded system, which provides two ungrounded (hot) wires and one grounded conductor that is a combined protective earth and neutral (PEN) conductor. At the service the PE or EGC and the neutral are separated and are maintained that way throughout the facility.

Utilities also provide 120/208, three-phase, four-wire wye-connected supply. This system provides three-phase power for loads such as motors at a nominal voltage of 208 V, while other loads are supplied with 120 V nominal. Utilities with 120/240-V, three-phase, four-wire delta-connected services provide three-phase power to loads such as motors at 240 V nominal, and other loads are serviced at 120 V nominal. Utilities at 277/480 V, three-phase, four-wire delta-connected serve three-phase loads at 480 V nominal, and loads such as lighting are served at 277 V nominal. This type of system does not provide for 120-V loads.

Industrial facilities can be provided with 2,400, three-wire delta, while 4,160, 12,470, and 34,500 V three-phase are four-wire wye-connected services.

12

ELECTRICAL LIGHT FIXTURES

The invention of the incandescent lamp was the "killer application" responsible for the emergence of electric utilities and the rapid build-out of the world's electrical infrastructure. Once reliable electrical energy supply was established, appliances such as fans, irons, vacuum cleaners, and refrigerators soon became household commodities. In the early1940s, the fluorescent lamp began to be applied to the workplace as a source of less glaring illumination. These lamps soon became the light source choice in commercial and industrial facilities. The next wave of change was fostered by the development of high-intensity discharge (HID) lamps. They were found to be well suited for applications with high ceilings. The current stage of lighting advancements began with the energy crisis of the 1970s. This stage has made use of electronics to reduce the operating cost of light fixtures.

Electronic ballasts allow for operation of fluorescent lamps at high frequencies (20 to 30 kHZ) and have resulted in an increase in lamp efficacy of some 25 percent and made it possible to dim fluorescent lamps over a wide range of light output. This, in turn, fostered the further development of energy-lighting management systems by other than simple on-off time clocks. Significant increases in lamp efficiency have allowed the Illumination Engineers Society (IES) to reduce recommended lighting levels by as much as 50 percent.

APPLICATIONS FOR LIGHT FIXTURES

The common uses of light fixtures are:

▨ Access and egress
▨ Facility security
▨ Task lighting
▨ Outdoor athletic fields
▨ Indoor sports lighting
▨ Roadways and traffic intersections
▨ Underpasses and parking lots
▨ High bay manufacturing
▨ Warehousing and retail
▨ Lay-in ceiling
▨ High ceiling
▨ Automotive garage
▨ Building outline lightning

DESIGN OF LIGHTING SYSTEMS

The design of lighting systems can be separated between indoor and outdoor illumination systems. Indoor lighting systems can be divided into lighting for general areas such as walkways, manufacturing, and machine or process lighting, while outdoor lighting is primarily directed toward area lighting such as entryways to facilities' parking and buildings. Interior lighting design is made more complex in that different surfaces reflect light at differing rates and different tasks require different levels of lighting.

Illumination is typically measured in units of foot candle (fc), or in Lux (lx). One foot candle is equal to 10.4 Lux. One foot candle is the amount of illuminance incident upon an area of one foot square. Low levels of illumination for sidewalks or parking lots are provided primarily for safety and trip hazards.

There have been several studies that show a high degree of correlation between personnel productivity and lighting levels. A lighting design must consider energy restrictions, contrast, and avoidance of washout producing glare. It is not unusual for the lighting level requirements to change during the workday or for various shifts. Building automation systems (BAS) and lighting management systems are used to optimize total building performance by providing only the level of lighting needed at the time, location, and nature of task and the presence of personnel. Modern office facilities tend to be designed so as to make effective use of natural outdoor ambient light.

MAJOR TYPES OF LAMPS USED IN COMMERCIAL AND INDUSTRIAL FACILITIES

The following are the major types of lamps used in commercial and industrial facilities:

- Incandescent
- Fluorescent
- High-intensity discharge (HID) mercury vapor metal halide
- Low- and high-pressure sodium

LIGHT FIXTURE PERFORMANCE

Sadly, the expectation level of the performance of lighting is so low that if the light is illuminating many think it is working satisfactorily. Figure 12–1 is a drawing of a common handheld light meter. A light meter is seldom used to properly aim a fixture so that the light is incident upon the desired surface. In a highly litigious society, light pollution and trespass lawsuits are increasingly being brought before judges. Fixtures installed outdoors should be dark sky or neighborhood friendly. The installation of cutoff fixtures should be a matter of common sense and not the result of costly legal litigation.

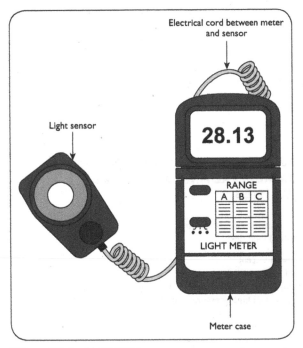

Figure 12-1 Lightmeter

PATTERN OF ILLUMINANCE

A photometric pattern is the pattern of illuminance produced by a fixture. The pattern shown in Figure 12–2 indicates how light is distributed across an area by a light fixture suspended high and aimed down. This pattern lists the amount of light incident upon an area in foot candles.

Figure 12–3 is a photometric pattern and indicates how light is distributed across a building wall by a light fixture installed at ground level

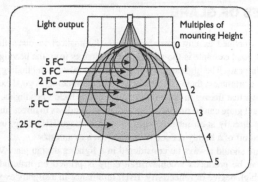

Figure 12-2 Photometric Pattern on A Flat Surface

and aimed up. To properly select and install a light fixture the desired photometric pattern needs to be determined and the correct light fixture selected and properly installed.

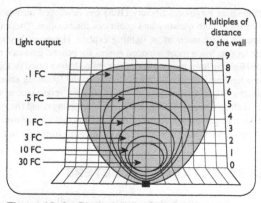

Figure 12-3 Photometric Pattern on a Wall

TYPES OF GLARE

Glare is defined as unwanted brightness (luminance) in one's field of view. A good example is the glare one experiences from the headlights of oncoming cars. The highest level of glare is described as a blinding glare. That is, a brightness that is so intense as to be debilitating. A good example is the deer that freezes in the blinding glare of a vehicle's headlights.

Direct glare can be produced by unshaded windows in direct sunlight, looking directly at an incandescent or high-wattage HID lamp, or light reflected off of a highly reflective surface such as a mirror.

Glare should always be considered in a lighting design plan. Veiling glare can be produced when written copy is printed on materials that have a high degree of reflectivity. Even luminance in a slightly excessive amount over time can produce a level of discomfort and annoyance to personnel. Discomfort glare has been known to result in eye fatigue and headaches. Reflective glare can result in a reduction of visual performance by personnel or an increase in the general level of discomfort.

ILLUMINATING ENGINEERS SOCIETY

The Illuminating Engineers Society (IES) has developed recommendations in terms of both quantity and quality of illumination. These values are the result of a consensus of lighting experts. However, they should be understood as being recommended ranges for general guidance only.

Today lighting level recommendations are no longer based on the idea that more lighting is better lighting. Generally the following should be factored into a lighting plan: age of the occupants; reflectance of major surfaces such as ceilings, floors, walls, and furnishings; and the type of visual task to be performed. Some of these factors may change over time. This results in the need to tune illumination systems from time to time.

LAMP SELECTION PARAMETERS

The key parameters by which the various types of light fixtures can be compared are:

- Lumens per watt
- Ballast efficiency
- Lamp life
- Lumen maintenance
- Color rendition
- Color rendition stability and control over the lamp(s) life

LIGHT FIXTURE OPERATING CONTROLS

Light fixtures can be controlled by the following methods:

- Remote or local switching
- Manual or automatic switching
- Motion detector
- Photo cell
- Time clock
- Time clock and photocell
- Relay in panelboard-load center as part of a building automation system

PHOTOCELLS

Figure 12–4 shows one of the various styles of photocells available. The viewing window is typically pointed north.

INCANDESCENT LAMPS

The incandescent lamp has the following characteristics: it is a continuous source of light, a point source, that can be dimmed over a large range. It has a low production cost and is temperature independent. Additionally, it has a small weight and size form factor. The most common base used on incandescent lamps is the Edison base. This type of lamp has the following negative factors: it has a low efficacy and an

Figure 12–4 Typical Removable Photocell

operating life of only 1,000 hours. It produces a source of light glare and its operating cost is considered to be high. Its low first cost will likely result in its continued use for many years to come. This family of lamps is available from 8 to 1,500 watts. Some are manufactured with the inside surface of the bulb wall frosted so as to provide a more diffused light distribution.

GAS DISCHARGE LAMPS

Fluorescent and high-intensity discharge lamps are gas discharge lamps. There are over 30 types of gas discharge lamps available today. The following are the more common ones:

▦ Ceramic metal halide lamp
▦ Compact fluorescent
▦ High-intensity discharge lamp
▦ Mercury vapor lamp (MV)
▦ Metal halide lamp (MH)
▦ Quartz metal halide probe-start

■ Quartz metal halide pulse-start
■ High-pressure sodium
■ Low-pressure sodium

FLUORESCENT LAMPS

A fluorescent lamp is a low-pressure mercury gas discharge lamp. This family of lamps has the following positive characteristics: they have a high efficacy with a long lamp life. The light output is of low intensity and diffused. The lamp has a negative feature in that it requires a ballast to change the AC supply to that required by the lamp. Additionally these lamps are considered to be bulky compared to other lamps. Initial cost is greater and the fluorescent lamp output is sensitive to ambient temperature changes. Sizes range from 7 to 100 watts. Lamps come in instant start, rapid start, and preheat types. They have a service life of 10,000 hours for preheat and instant start lamps, while rapid start lamps are around 20,000 hours. Compact fluorescent lamps (CFLs) are rapidly replacing incandescent lamps. CFLs are four times as efficient as incandescent lamps and their service lives are in the area of 10,000 hours compared to 1,000 hours for the incandescent lamps. The inner walls of fluorescent lamps are coated with a fluorescent phosphor that fluoresces when in operation. The ballast supplies the high voltage required to start the lamp and then to control the current drawn by the lamp.

FLUORESCENT BALLAST

All fluorescent lamps use a ballast to raise the voltage and to control the current flow. The ballast specifications differ for each type of lamp and are specified by an American National Standards Institute (ANSI) standard. Magnetic ballasts are made to supply one to three lamps. Electronic ballasts are made to supply one to four lamps. Electronic ballasts allow for dimming over a wide range without reducing lamp operating life. With adequate filtering harmonic currents reflected back to the power supply can be reduced to around 10 percent with power factors close to one.

HIGH-INTENSITY DISCHARGE LAMPS

A high-intensity discharge lamp is a high-pressure gas discharge lamp. This family of lamps is considered to be an intense source of light at high levels. It is considered to be a point source of light and operates at a high efficacy level. HID lamps use a screw in base, much like the Edison, only about two times larger called a *mogul base*. Like fluorescent lamps, HID lamps require the use of a ballast. They are not sensitive to changes in ambient temperature. Their high first cost and an amount of cold start and restrike delay in developing illumination are also negative factors. The most common types of HID lamps are mercury vapor, metal halide, and high-pressure sodium lamps.

MERCURY VAPOR LAMPS

Mercury lamps are filled with mercury gas at a pressure of between two and four bar (standard atmospheric pressure). The inner walls of these lamps are coated with a phosphor coating to improve color rendering. Lamp sizes range from 40 to 1,000 watts. Operating lives are around 24,000 hours. Use of this type of lamp is declining due to its low efficacy.

METAL HALIDE LAMPS

These types of lamps use one or more metal halides (bonding between halogen and the metal) such as sodium, thallium, indium, and dysposium. Metal halide lamp sizes range from 40 to 3,000 watts. Service life is around 12,000 hours. These lamps provide better color rendering than mercury vapor lamps.

HIGH-PRESSURE SODIUM LAMPS

The high-pressure sodium lamp contains a mixture of mercury, argon, and sodium. Lamp sizes range from 40 to 1,000 watts. Service life is typically 24,000 hours. Color rendition is poor, thus allowing for use in applications such as roadway illumination and parking lots.

TYPES OF HID BALLASTS

The following is a brief list of the various types of ballasts:

▪ Low-frequency magnetic regulation (standard ballast)
▪ One-coil reactor (R)
▪ Two-coil high-reactance autotransformer (HX)
▪ Constant wattage (CWA) autotransformer (isolated)
▪ Three-coil regulated lag (REG LAG), also known as magnetic regulated (MAG REG)
▪ Electronic high frequency
▪ Combination of electromagnetic and electronic
▪ Vacuum impregnated for noise reduction
▪ Standard open construction
▪ Encapsulated-potted using asphalt or polyester and silica

WHAT BALLASTS DO

A ballast provides the voltage and regulates the current required by the lamp to operate. When the ballast is not designed to provide the voltage required to start the lamp, an *ignitor* or *starter* is added to the *secondary side* of the ballast. To improve the power factor of a ballast, a *capacitor* can be installed electrically parallel to the *primary* windings. A capacitor added to the secondary side limits the current flow to the lamp. A capacitor connected to a separate secondary coil will improve the ballast's ability to regulate voltage to the lamp.

BALLAST VOLTAGE RATINGS

Fluorescent and HID ballasts are commonly available for use with the following building supply (*nominal*) voltages:

- 120 VAC
- 277 VAC
- 120/277 = dual tap
- 120/208/240/277 = quad tap
- 120/208/240/277/480 = five tap

Ballast Voltage Taps

The *primary (line-supply)* side of a ballast can be provided with several taps (leads connections) for use with more than one voltage. Only two connections are made to the primary side of the ballast at one time, those being *common* (neutral-white insulation) and *line connection (L-1)*. The common connection is used as the line two (L-2) connection. When the nominal voltage is either 120 or 277 VAC, one connection will be the common (neutral) and the other will be the L-1 connection.

BALLAST WIRING DIAGRAMS

Chapter 19 contains several diagrams showing how various types of ballasts are wired.

13

ELECTRICAL TIMERS

This chapter covers the following items about electrical timers:

- The five sections of timers
- Nine timer operating principles
- Five common modes of timer operations
- Timer charts

SECTIONS OF A TIMER

Figure 13–1 provides a block diagram that shows the five functional sections of a timer. Timers can be divided into the following sections:

- Power or energy input
- Signal input
- Adjustment-mode selection
- Decision-timing mode
- Output

Energy Input Section

The energy input section receives the electrical energy from a supply source. The needed energy could also come from a mechanical source such as a wound spring.

Signal Input (Start-Initiate) Section

The start, the trigger, or initiate signal serve the same purpose, they connect an external signal to the internal decision-making section of the

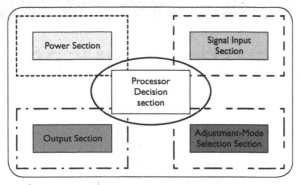

Figure 13–1 Timer Block Diagram

timer. A timer may or may not have an external start input. The timer could be built so that when power is applied, it automatically begins its timing function.

Adjustment-Mode Selection Section

Some timers can be adjusted for the amount of time, or the mode of operation, or both.

Decision-Timing Mode Section

In this section, the timer receives a signal and compares it to its instructions program, makes a decision, and then takes some action, such as energize an output relay. The decision section receives information (in the form of an electrical signal) from the input section, and commands the output section to take some action.

Output Section

This section interfaces with the electrical circuit of the machine controlled by the timer. This is the section that controls both the instant (when provided) and the time-delayed contacts.

Other Section Descriptions

Timers can also be broken down into the following sections:

■ Timing means: Determines the range and accuracy of the timer.
■ Starting means: Initiates the timing; can be either automatic or manual.
■ Setting and control: Actuates the controlled contacts; can be adjusted for desired time intervals, and possibly modes.
■ Load contacts: Links the timer to the controlled process. May switch a different voltage than the timer.
■ Indicating means: May be provided to indicate the status of some function of the timer.

FIVE COMMON MODES, OR FUNCTIONS, OF TIMERS

First let us build a level of understanding about the five basic types of timer modes or functions. Timers can accomplish the following:

■ Compress an event; make it occur in less time.
■ Expand an event; make it occur over a longer time.
■ Delay, or shift an event; make it happen at some point in the future.
■ Make a single event occur repeatedly.
■ Cause a sequence of events to occur once or repeatedly.

The five most common timer modes of operation are:

■ On-delay
■ Off-delay
■ Single shot
■ Repeat cycle
■ Interval

There is no national standard for the names of the various timer modes. This results in many different names used for timers that accomplish the same task. For example, on-delay timers are also known by the following terms: delay-on-make timer, delay-on-operate timer, delay-on pick-up, delay-on-energize, and sadly, by many other terms.

Starting a Timer's Mode of Operation

A manufacturer can use any one of the following methods to start a timer:

▦ Apply power to the timer
▦ Apply-remove a signal to the timer's initiate terminal
▦ Apply a continuous signal to the initiate terminal
▦ Remove a signal from the timer's initiate terminal

Depending upon the specific timer, applying or removing the initiate signal may or may not result in the timer resetting. That is, once applied, the initiate signal may or may not be ignored during the timing period.

NINE TYPES OF TIMER OPERATING PRINCIPLES

The following is a list of the various operating principles used by timer manufacturers:

▦ Spring wound
▦ Motor and gears (time clocks)
▦ Electrothermal
▦ Electromagnetic
▦ Air-pneumatic
▦ Electromagnetic and hydraulic
▦ Electromechanical
▦ Solid state
▦ Photoelectric

Solid State, Microprocessor

Today the workhorses of industry are the logic timers, used in programmable logic controllers (PLC), and the solid state (SS) electronic,

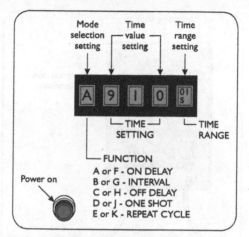

Figure 13–2 Solid State Multimode Programmable Timer with Digital Display and Power on LED

integrated circuit (IC) microprocessor-based microcontrollers. These timers are reliable, small, low cost, multifaceted, and field adjustable. Figure 13–2 is a drawing of a typical multifunction timers face panel, while Figure 13–3 is a drawing of a simple two wire single-function adjustable timer. These types of timers are capable of switching faster than an electrical relay and can communicate with plant-wide controls systems such as field bus and other digitally based control systems.

Photoelectric Timer

The photoelectric timer uses an ambient light–powered switch. Light as energy is converted to electrical energy by a photocell. A photocell produces a very small amount of electrical energy; it is enough to operate a small relay. Many times, outdoor security lights use a small photocell timer to turn the light fixture on and off at specific levels of area luminance.

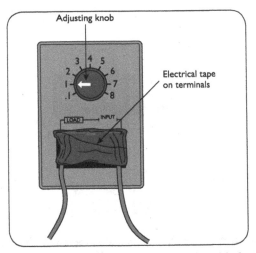

Figure 13–3 Solid State Two-Wire Timer, Adjustable from
One to Five Minutes "Cube" Timer

On-Delay Mode Timer

An on-delay timer begins its timing period when power is first applied
to it. Its controlled contacts will change position when the timing period
has elapsed. When power is applied to the timer, the timing period
begins. Should power be removed, the timer resets to zero time. When
power is reapplied, it begins the timing period from time zero. This will
keep a controlled load from briefly turning on and then off, or short
cycling.

Off-Delay Timer

An off-delay timer is used when one portion of a machine needs to keep
operating after another has stopped. The delay period begins when the
timer is turned off. Upon completion of the time period the controlled

contacts change position. The off-delay timer is also known by the following terms: delay-off timer, delay-on release, delay-on dropout, post-purge delay, and power-off timer. Some timers do and some do not require a separate signal to begin-initiate-start their timing operations. Almost all timers require electrical power for them to operate. All timers will have a time delay relay (or solid-state equivalent) that controls the load. When looking at a timing chart, if a start-initiate signal line is not seen, the timer will start when power is applied or removed from the timer. When looking at a timing chart and an initiate or start line is seen, the timer will begin its timing operation when a signal is received or removed from that terminal on the timer.

Interval On Mode Timer

The words "interval on" indicate that the output relay's normally open contact will close immediately when power is applied to the timer. Interval on timers are also known as on intervals, pulse shaping timers, single pulse on operate, and bypass timers.

TIMING CHART LEVELS

Timing charts are graphs of the states of individual elements of a timer, such as power, initiate-control signal, and controlled (both instant and time delayed) contacts. Each of these elements is represented by a line. Time zero is on the left and time increases to the right. The top line of a timing chart always represents the power to the timer. The second line is for the separate reset function when it is provided. The third is for the gate function. The fourth line is for the N.C. When a set of electrical contacts will pass power they are said to be closed (C). When they are closed with no voltage applied to the relays coil, they are considered to be normally closed contacts (NC). The fifth level line is for N.O. When a set of electrical contacts are open they will not pass power. The sixth level line is for an LED when one is provided.

Timing charts have only the lines required by the specific timer represented by the timing chart.

Top Level Operating Power

Timers require some type of energy input to operate. The top line on a timing chart represents this power to the timer. It is common practice to think of the brains of a timer as the equivalent of an electromagnetic coil that controls the relay contacts even when the timer uses solid state components.

The words "power to the coil" are used in this work, even though the timer may not have an electromagnetic coil. As a timer can or cannot have power to it, a timing chart indicates both conditions. Figure 13–4 is of the basic symbol used to indicate the status of various signals in a timer. A horizontal line is used to indicate the state of power to the timer. A vertical line above the horizontal line indicates that power is applied to the timer. A second horizontal line to the right indicates the passage of time. A second vertical line below the time horizontal line indicates that power has been removed. The second vertical line is followed by yet another horizontal line indicating the passage of time. The space between the two vertical lines indicates the passage of time. The horizontal lines in a timing chart are typically not marked in units of time such as seconds, minutes, or hours.

Start-Initiate Signal Level

Some timers have an internal jumper that passes power to initiate the timer's action; others have a separate initiate (or start) terminal that must receive an external electrical signal to begin the timing operation. Still others use external "dry contacts," where no foreign voltage is required.

The closing of these contacts starts the timing operation. The second level on a timing chart by convention is used to graphically

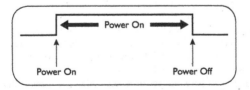

Figure 13–4 Timing Chart Power Level Diagram

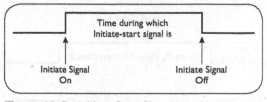

Figure 13–5 Initiate-Start Signal Level

show if the start or initiate signal is on or off, the same way power is shown to be present-absent in the first line. Figure 13–5 shows initiate signal in the both the On and Off (power passing and not passing) states. The width of the horizontal space between the two vertical lines indicates the amount of time that has passed. The initiate, or control, signal may or may not need to be present during the timer's operating mode. With some timers, removing the initiate or control signal will cause the timer to "reset." With others, removal of the signal has no impact. For still other timers, removal of the start signal stops time; it freezes it at its current value. When the signal returns, the timer starts at the exact place it previously stopped, and then times out. The word "initiate" is used to indicate the starting of the timing action. With some timers, after the initiate switch has been closed, opening it will have no effect on the timer. Thus while the initiate switch may start the timer, it may not be able to stop it. That is, the initiate signal may be ignored by the timer. The words "start" and "initiate" can be understood as the same. However, initiate and control cannot be considered to be the same.

Instant and Time-Delayed Contacts Levels

As an optional feature, in addition to the time-delayed contacts, a timer may also have a set of instant or non-delay contacts. As the timer controls both sets of contacts, they could be called controlled contacts. Care should be exercised to communicate what is intended by using specific words. Thus, the words "controlled" contacts and "time-delay" contacts could be used. The timer controls both sets, one is non-delayed, and the

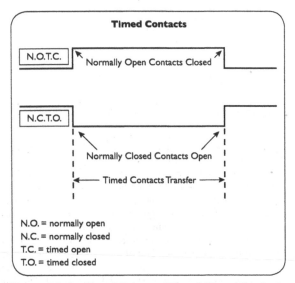

Figure 13–6 Timed Open and Timed Closed Contacts

other is subject to some amount of time delay. Figure 13–6 provides a drawing of both options of a timers contacts.

Timers can have more than one set of time-delay contacts. A timing chart will have one line to indicate the state of the timed contacts. It may also have another line indicating the position of the instant non-delay contacts when provided. It can be confusing when talking about normally closed contacts opening, and normally open contacts closing. It would be less confusing if the contacts were said to change position or change state, or transfer. Instant contacts change position immediately when power is applied and again instantly when power is removed. The timed contacts are going to change position (transfer) when the timer's mode of operation calls for them to change position.

The letters T.C. indicate timed closed, and T.O. indicate timed open. Both initials are sometimes used on timing charts.

Additional Timing Chart Levels

Some timers may have an inhibit signal line. This input stops the timer's timing. When it is removed, the timer starts adding to the previous value. Some timers may have a line for a reset signal. When the reset signal is applied, the timer resets so that it begins a completely new timing operation.

COMPLETE TIMING CHARTS

The discussion of timers has progressed in small steps, and an example of a complete timing chart has not been provided. Figure 13–7 provides an example of what a typical off-delay timing chart might look like. The chart has three levels that represent the power, initiate, and timed contacts. In the top level, power is applied to the timer. The

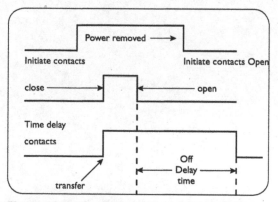

Figure 13–7 Complete Off Delay Timing Chart

timed contacts do not change state. When the initiate contacts pass power to the timer, the timed contacts remain as they normally were. Note that when the initiate contacts open, the timed contacts still do not transfer. It is only when the amount of preset time has passed that the time-delay contacts transfer. Notice that even though power has been removed from the timer, the time-delay contacts do not transfer. It is only when the preset amount of time after the initiate contacts have opened do the time-delayed contacts transfer. That is why it is called an off-delay timer. When the initiate signal was removed, the timer began to "time out." When the preset time expired, the time-delay contacts transfer. There are many types of timers, but all timers use the same type of timing chart to communicate their action.

14

ELECTRICAL DIAGRAMS

This chapter covers ladder and one-line type diagrams.

ELECTRICAL LADDER DIAGRAMS

As it is difficult to remember where every wire in every machine is connected, electrical diagrams are very important. Electrical ladder diagrams replace wires with lines and parts with symbols that communicate how the machine operates. No single electrical diagram will provide all of the information needed to understand and troubleshoot a power distribution system or even one small machine. Diagrams have title blocks that provide limited information, such as the plant facility, or machine name, date of the drawing, and a few other details. A list of symbols may also be provided. Chapter 19 provides a list of various symbols used in both types of these diagrams. Review of these symbols will when determining what a symbol means before it is encountered in electrical diagrams.

ELEMENTS IN ELECTRICAL CIRCUITS

Figure 14–1 provides the framework of a typical control circuit in ladder diagram format. Electrical circuits and their symbolic diagrams typically contain the following elements:

- Source of power
- Path for current flow

▦   Load to consume the power
▦   A control device to start and stop the load
▦   A protective device, such as a fuse or circuit breaker

The electrical wiring in a machine and a ladder diagram look very different. Lines are symbols used to represent the real wires, while various symbols are used as stand-ins for actual parts.

Figure 14–2 is composed of two diagrams. The upper part is the power diagram. The ladder diagram of the pilot duty control devices is in the lower portion. The control devices are the Stop and Start buttons, the phase monitor relay, the motor starter coil, and the overloads. The main power portion contains the source L-1, -2, and 3-, the motor starter contacts (M), the overloads (OL), and the three-phase motor.

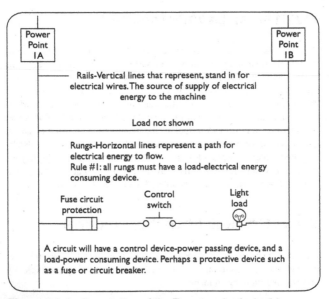

Figure 14–1   Description of the Framework of a Ladder Diagram

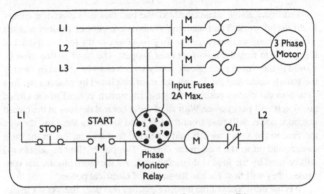

Figure 14–2 Power and Control Schematic Diagram

Figure 14–2 uses symbols, and the machine uses real parts that do not look like the symbols.

Because the diagram and the physical machine look very different, it is important to study the symbols used in diagrams before attempting to read electrical diagrams and troubleshoot the machines.

Figure 14–3 contains the source of power and two fuses (the two rectangular boxes).

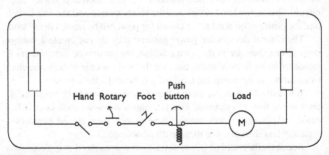

Figure 14–3 Single-Rung Ladder Diagram

The figure also contains the load, the part that uses electrical energy to do useful work. The remaining symbols are control devices and the wires that connect them together. To pass power to the load so it can use that power to run, the following must happen: The hand switch must be moved to the closed position, the rotary switch must be closed by turning the switch knob, and the foot switch must be closed by pushing the foot switch to the closed position. As the push switch is held in the closed position it will pass power. With all of these control devices in the closed position, power will pass from the left-hand source to the fuse, through the fuse to the rotary switch, and on to the foot switch, out to the push switch, and on to the load in the circuit. The power (flowing electrons) will be used by the load to do useful work. As the electrons are not consumed, they will flow back to the source of electrical power.

It is the movement of the unseen electrons that does the work. Physics tell us that electrons can neither be created nor destroyed, that is why they come from the source, through the protection, control, load, and back to the source. For the machine to operate, the electrons must flow into and out of each wire, fuse, switch, load, and return to the source. If any one of the fuses, control devices, hand switch, rotary switch, or foot switch or push buttons opens, power will not be passed to the load. Thus the load cannot do useful work, that is, it will not run. By tracing the flow of the electrons to and through each part of the circuit one can understand how the machine works. In this diagram there are four control devices connected in series. The first three are normally open, and the fourth is normally closed. This electrical path could be described as being an "and" circuit. That is, the hand switch *and* the rotary switch *and* the foot switch *and* the push button must all be closed for power to be passed to the load.

The control devices are power passers; they do not create a voltage drop. When they are in the open position, the resistance across them is measured in millions of ohms. Because this resistance is so high, the electrons cannot jump across the air-filled gap between the contacts.

When the controls are closed, they create an extremely small resistance to the flow of electrons. That is, control elements such as switches cause only a very small voltage drop across their closed contacts—typically less than one-one thousandth of one volt.

Recall that a load is a power consumer; it uses the flowing electrons to do useful work. A load is designed to cause a voltage drop, the same as the circuit voltage. The ability to read ladder diagrams can be likened to working a set of muscles in a gym. The more the muscles are exercised, the stronger they become.

Let us now read another diagram shown in Figure 14–4. The horizontal line 1 running from the left-hand vertical line (rail) is called a rung. It begins and ends at the two rails. The horizontal line 2 does not begin at the rail, so it is called a run.

Following the path across rung 1, the top half of a Hand-Off-Auto (H-O-A) is seen. The second half of this switch is shown on run 2. Both halves are linked together and move at the same time. As shown, power passes through the control to the load, and as the overloads (OL) are closed, through them and back to the source. The load is identified as being a pump motor marked M-2.

Notice that the H-O-A contacts in run 2 are in the open position, so power is not passed to the open float switch. When the H-O-A is paced in the Auto (A) position, the contacts in rung 1 are opened, and the contacts in run 2 are closed, passing power to one side of the float switch. The float switch closes the circuit when the measured fluid falls below some point. That is, it makes its electrical contacts (they close passing power) on the fall of the fluid being measured. When the H-O-A is placed in the Off (O) position, both sets on contacts on rung 1 and run 2 are open, so no power is passed to the controlled load.

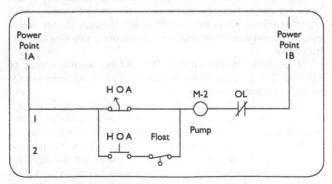

Figure 14–4   HOA Ladder Diagram

Let's review the diagram in Figure 14–2 again in more detail. Power flows to and through the Stop button as it is spring-loaded and held in the closed position, to the Start button, which has a spring holding it in the open position. When the Start button is pressed, it closes passing power to the load, M and as the OLs are closed, back to the source. The load M creates a magnetic field that draws the three main power contacts to the closed position.

When the main power (M) contacts close, power is passed to the over-load heater elements and then to the three-phase motor, and the motor runs. The phase monitor relay has to have voltage to all three of its terminals, numbers 3, 4, and 5, for its contacts, 1 and 8, to be closed so as to allow it to pass power to the M coil.

In order for the motor to remain online, the Start button must be held in the closed (pushed in) position. Notice the M contacts below the Start button—these contacts are physically linked to the main power (M) contacts. When the M contacts closed, the M contact below the Start button also closed, forming a parallel path for power to flow around the Start button when it is released.

This path keeps power flowing to the M coil and that keeps the M contact closed passing power to the three-phase motor. Loss of power from the source—a blown fuse, motor overload, or by someone pushing the Stop button—can all cause the motor to stop running.

All of the controls are in a portion of the circuit that has only a small amount of current flowing in it. Therefore, the control devices are called *pilot duty* devices. They carry a small amount of current and are used to control the operation of the equipment, much like a ship's captain controls a ship.

Many motors operate on 460 VAC, and the controls operate on 120 VAC. This is accomplished by locating a step-down transformer in the enclosure with the motor starter.

In Figure 14–5, a control circuit transformer can be seen on rungs 5 and 6. Notice that the primary is protected by two fuses. Notice also that one control device is a pressure switch that makes on a rise in pressure.

Notice in rungs 7 and 8 that both the H-O-A and the pressure switch can pass power to both the motor starter and the small red panel light. That means that two or more loads can be served by one control device.

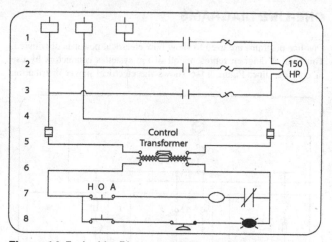

Figure 14-5 Ladder Diagram with Control Circuit Transformer

When the H-O-A is placed in the Hand position, and the pressure switch is open, power will be passed to both the motor starter and the red panel lamp.

When the H-O-A is placed in the Auto position, when the pressure switch closes, power is passed to both the lamp and the motor starter. The panel lamp is classified as an annunciation device. By studying this diagram, one can determine that if the motor is not operating, one of the following could be the considered likely causes:

■ No power to the machine
■ Motor overloads tripped
■ One or more of the fuses on the transformer are open
■ The pressure switch is open
■ The H-O-A switch is in the Off position

Ladder diagrams are vital tools used by electrical troubleshooters. They are quick and easy to read and provide a visual representation of how electrically powered equipment operates.

ONE-LINE DIAGRAMS

One-line diagrams are used to show how electrical power is distributed. This type of diagram represents all of the system's conductors by use of only one line. Figure 14–6 shows the electrical power distribution

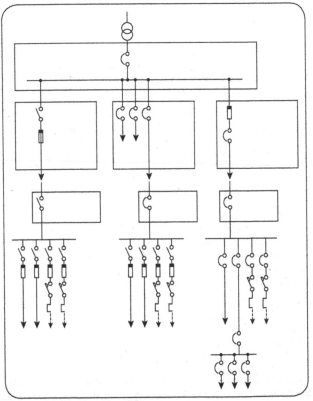

Figure 14–6 One-Line Diagram

in a manufacturing facility. The diagram shows that power is provided by the utility at 13.8 kV to a fused disconnect and then to a 1,500 kVA transformer. The transformer is connected in a delta primary pattern and grounded wye on the 480 three-phase secondary pattern. A step down power transformer (PT) provides 120 for voltage and watt-hour billing meter and amperage meter through 3,000:5 current transformers. A 3,000 A circuit breaker protects the 480/277 three-phase four-wire bus. Circuit breaker 2 provides power to a warehouse, while breaker 4 feeds power to the truck dock area. The last breaker, 6, provides power to cooler 1, refrigeration equipment. The wires providing power to the 1 cooler are three 3/0 and one #6 for the ground.

The symbol used for the circuit breakers indicates that they are draw-out types. Breakers 2, 4, and 6 are provided with ground fault protection of the equipment as noted by the letters GF in the circles. Chapter 19 provides a list of several of the more common symbols used in one-line diagrams.

15

ELECTRICAL TEST INSTRUMENTS

The three most commonly used electrical test instruments are: the clamp-on amp meter, the AC voltmeter, and the Ohmmeter. These three meters are commonly known today as digital multimeters (DMM). It is vitally important that electrical personnel know how to correctly use a DMM. These meters are used to not only understand how an electrical system is functioning, they are also very important to the safety of personnel working on and around electrical equipment. Figure 15–1 is of a common DMM. Today digital multimeters of all types are the most commonly used, so only they will be covered in this chapter.

WORDS USED WITH METERS

Much of the time meters can be used for various tasks, such as measuring AC and DC voltage, AC and DC amperage, and electrical resistance of a circuit. For this reason most of the time a meter will actually be a multifunctional measuring instrument. Some call them voltmeters, some multimeters or DMM, while others will simply say meter.

In this chapter, "multimeter" is used for a meter that can measure AC or DC voltage and electrical resistance, which is measured in ohms. When speaking about a meter that can measure amperage, "amp meter" will be used to indicate a clamp-on type meter. When a meter can measure resistances in the millions of ohms, at DC test voltages of 250, 500, or 1,000 VDC, "meg-ohm meter" is used.

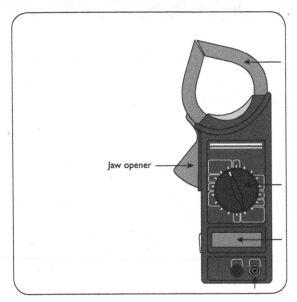

Jaw opener ───▶

Figure 15–1   Multimeter Diagram

Many of the modern family of electrical test instruments can perform additional measurement tasks in addition to the ones that will be covered in this chapter.

MULTIMETERS

As electrons cannot be seen, a voltmeter is needed to measure the pressure in an electrical circuit in units of measurement called volts (V). For this chapter, only alternating current voltages (VAC) will be considered. While many can, not all meters can measure direct current (DC) voltages. On some meters the letters AC and DC are preceded by the letter V on the meter's function selector scale. A tilde is used on some meters to indicate AC voltages. Some meters will use a solid line with three or more dots below it to indicate DC voltage function.

Voltage Function

Of all the functions electrical meters can perform, the most important is the measurement of VAC, as it is the most important to the safety of personnel. The maximum voltage of both the meter and its test leads must not be exceeded. While some cannot, many multimeters can read up to 600 VAC. Check the meter before it is used, as it could be destroyed by applying it to a voltage outside its design limits.

AUTO RANGING METERS

A modern DMM can be confusing and even a bit intimidating on occasion, even to old pros, with all of its buttons, knobs, and such. Figure 15–2 shows what I call the function selector switch found on a typical DMM. Some of the better quality meters have the ability to automatically measure

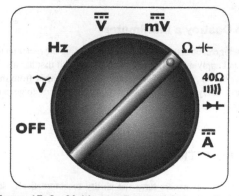

Figure 15–2 Multimeter Function Selector Switch

volts and millivolts. A millivolt is one, one-thousandth (1/1,000) of one volt. These meters will many times automatically select the correct voltage range using either a volts or millivolts scale. This can lead to confusion when one is not accustomed to this feature. This makes it important to carefully look at the meter's display to confirm the range being used.

For example, a machine is being checked to see if it has a voltage before being serviced, and the multimeter's test leads are connected to two test points, and the meter displays 87. This only could be 87 millivolts. When the letters mV are displayed, this indicates that there is essentially only a static voltage present. When an AC voltage is applied to a meter set to the DC voltage function, it may not display any value, or it may do something strange. Make sure the meter is set to the proper function as it can be destroyed in an instant.

THREE-STEP METHOD

All man-made things can fail, delicate instruments can fail internally without any indication on the outside. Always practice the three-step method when checking for voltage: Step 1, test the meter on a known life circuit. Step 2, test the circuit to be tested. Step 3, test the meter again on a known live circuit.

How to Destroy a Multimeter

A multimeter can be instantly destroyed by placing the meter on the ohms function and applying the two test leads to a circuit that has a voltage. In an instant, *boom, pop*, all of the magic smoke escapes from the meter. This is most likely to occur when one is under pressure to get a machine up and running.

AUTO FUNCTION FEATURES

Some of the better meters have the ability to automatically select the desired function, such as measuring volts or ohms. These meters have only an on-off selector switch. When turned on and the two test leads are applied to the circuit, the "internal brain" of the multimeter automatically selects the proper function, volts or ohms, and the correct range.

While battery life of multimeters is good, at some point they will discharge. Place a spare battery in the case with the multimeter as the battery will likely fail when it is needed most.

Leather Gloves

Find an electrician, and ask if he or she has ever been shocked. Then ask if he or she has ever been shocked while wearing a pair of leather work gloves. Some electricians do not always follow the safe work practice of using double-insulated gloves. Some companies do not provide them. Make it a lifesaving good habit of putting on at least a pair of leather work gloves before using a multimeter. Use these gloves only when using a multimeter. Special double-insulated electricians' gloves are always the best choice for protection from electrical shock. Leather work gloves without any metal trim are better than exposing bare skin to potential shock hazards.

OHMMETERS

This function of a multimeter measures the opposition to the flow of electrons in units of measurement called ohms. An upside down letter u, or a horseshoe, is used to indicate the ohm function on a meter. One of the rules of electricity requires the current flow (measured in amperes, or amps—A) to decrease when the resistance increases. An ohmmeter is used to measure this resistance. Before an ohmmeter is connected to any circuit, there must not be any voltage present. Capacitors (small canisters that act like storage batteries) must be discharged. First, test to confirm that there is no voltage present. If a reading other than the system's rated voltage is found, inspect the system for the presence of a capacitor. When one is located, it must be safely discharged. This can be done by shorting one capacitor terminal to the other capacitor terminal. Be careful, as there will be a small spark and a *pop* sound. Use only an insulated section of electrical wire for this task. Ohmmeters have ranges over which they will read resistances accurately. Common scales are 200, 2,000, 20,000 (20K), 200K, and 2,000K (2 meg-ohms). When the meter is set on the 200-ohm scale, and is used to measure 250-ohm resistor, it may display the letters OL, or the number one. Both are used to indicate that the measured value is in excess of the scale setting.

Ohmmeters are used to determine the amount of resistance between two points, for example, the two wires of an extension cord. While these insulated wires are physically touching each other, they should not be electrically touching each other. The resistance of many types of modern electrical insulation is about 150 volts per mill (1/1,000 of one inch) of insulation thickness. It is not uncommon for individual electrical wires to be insulated with 15 or more mills of insulation covering them. When an ohmmeter is used to measure the resistance of insulated wires in an extension cord, the resistance should be in the millions of ohms. Over time all electrical insulation ages and loses some of its electrical insulating ability, that is, its dielectric strength decreases. This is just part of the normal aging process. When a defect in electrical insulation occurs in one small area, electrons will begin to leak through the defect, allowing more and more current to flow. This is known as a short circuit.

The ohmmeter function of a multimeter is used to locate short circuits in electrical devices such as wires, motors, and transformers. A short to ground can be located by placing one test lead on case ground (any metal part of the machine) and placing the other test lead to each of the suspected wires one at a time. As a general rule a value less than 1 million ohms indicates that the conductor is shorted to ground.

If all of the insulated wires are healthy, they will tend to read over 1 million ohms. The one wire that indicates a very low resistance value is most likely the cause of the trouble and should be investigated further.

Three-phase motors draw electrical energy from the three power supply wires equally. So the resistance between each should be very close to the same value. Using the ohm function of a multimeter, a measurement from each of the motor's leads (wires) to each other, the meter should display the same value. Any reading of more than 5 percent difference should be investigated. When a load such as a motor has a winding that has shorted open, the resistance will be over 1 million ohms. When a load such as a transformer has shorted a winding to itself, the resistance will be very low, typically less than one ohm.

MEG-OHMMETER

A meg-ohmmeter (known as a "megger") is a basic ohmmeter with the added ability to measure resistances in the millions of ohms. A meg is equal to 1 million. Meggers typically can measure resistances as high as 1,000 or more meg-ohms, whereas the typical ohmmeter can measure only values up to about 5 meg-ohms. There are minimum recommended values published by organizations such as the National Electrical Manufacturers Association (NEMA) and the International Electrical Testing Association (NETA). These minimum values have some variables associated with them, along with the specific type of insulation resistance test that is to be conducted.

For many years a value of two meg-ohms for a 460-V motor tested at 1,000 VDC has been used as a go-no-go type rule of thumb. Never test any machine that has some type of solid state device in it with a megger. To do so will instantly destroy these sensitive components. Disconnect all electronic devices before "megging" any circuit. Meggers typically have selectable test voltages of 250, 500, 1,000 and some as high as 4,000 VDC.

CLAMP-ON AMP METERS

When current flows over an electrical wire, a magnetic field is produced. That is, the wire becomes electromagnetic.

The intensity strength of this field increases as the amount of current flow increases. An amp meter is used to measure this magnetic field without being electrically connected to the circuit being tested. Clamp-on amp meters are opened and then placed around one wire and then closed. The meter senses the strength of the magnetic field and displays the measured value on the meters display in amperes (amps).

Electrical motors typically have an amperage value listed on the motor's nameplate. This may be listed as either RLA, for rated load amps, or FLA for full load amps, both being the same value.

When a three-phase motor is fully loaded, it will pull full-load amps. When the motor is overloaded, it will pull more than rated load amps. And when it is unloaded it will pull less than full-load amps.

Clamp-on amp meters, like all test equipment, are used to provide answers to the electrician's questions. For example, is this motor fully loaded or overloaded? Are all three phases of this motor pulling the same amount of current? The amp meter can quickly be used to obtain answers to these important questions.

There are lots of different brands and models of multimeters, Meggers, and clamp-on amp meters available today. Some sell for only a few dollars, others for thousands of dollars. When selecting a meter, look for features that will be routinely used, such as AC volts to 1,000 AC amps to 1,000 resistance values to 5 meg-ohms. Then consider the category of the meter.

For general commercial and industrial work, where 460 V and short-circuit amperages in the 1,000s will be encountered, a Category III or higher meter should be selected. Electricians routinely check a machine for the presence of a voltage. Often, when they say there is no voltage present, they feel it is safe to touch the circuit's wires and components bare-handed. Electrical test instruments should be selected as one would any piece of life safety equipment.

16

ELECTRICAL TROUBLESHOOTING

In the 1950s some electricians would troubleshoot by pushing a relay in to see if the machine would run. Increased use of solid state components such as PLCs, timers, and VFDs has resulted in fewer "stuff" to push. Today machines are more complex, requiring the use of diagrams, DMMs, documentation, and on occasion, a laptop computer.

INTRODUCTION TO TROUBLESHOOTING

Troubleshooting should begin before the machine fails. It is best to prepare for success and avoid failure. The encouragement to "plan your work and work your plan" is of value. This chapter provides a sampling of information that will help improve one's troubleshooting effectiveness.

Some act as though restarting the machine has somehow fixed it. "Circuit breaker tripped" is a good example. The troubleshooter resets the breaker, restarts the machine, considers the trouble to be a "nuisance trip" and proceeds to the next trouble call. The amount of effort invested in troubleshooting was almost zero. Correcting the root cause of the failure requires more than replacing the broken part. Good troubleshooting involves the following steps:

- Preparation
- Planning
- Following proper procedures
- Following correct machine repair and start-up procedures

▨ Using proper test and measurement instruments

▨ Using modern analytical software methods and equipment

▨ Reviewing all of the above with an eye toward potential improvements

NOT WHERE IT IS, BUT WHERE IT WAS LOST

Follow along as a common trouble is described, to build support of the statement "not where voltage is, but where it was lost." For example, a maintenance hand needs to drill a hole, so he gets a 100-foot extension cord, a GFCI, and drill out of the truck. He goes to the location where he needs to drill the hole and plays the cord out from the work location to a distant 120-V outlet. He plugs the cord into the wall outlet and goes back to the work location, plugs the drill into the GFCI, and the GFCI into the cord, and starts to drill the hole. Only the drill will not run. He resets the GFCI and the drill still will not run. Using a DMM he determines that no voltage is present at the GFCI. The end of the extension cord is checked, and there is no voltage there either. There is 120 V present at the wall outlet when it is tested. The problem is voltage has been lost between the two locations, where power is and where it is needed. The trouble is not with the outlet, as voltage is present there. The trouble is located in the 100-foot-long extension cord. The drill was used earlier in the day and it worked fine, and the GFCI was checked, and the end of the extension cord was checked.

The extension cord has been determined to be the cause of the trouble. A visual inspection of the extension cord reveals that a heavy steel I-beam was dropped on the cord and cut one wire in two. The drill did not run because voltage was lost at one point in the extension cord. That is, in a small section of the cord, on one side there was voltage, and on the other side the voltage was not present.

This important point is lost to some technicians. They check for the presence of a voltage. They fail to understand that it is the loss of voltage that keeps a machine from running. Find the two closest points where voltage is and where it is not, and the trouble will have been pinpointed.

Given that the machine worked in the past, it is logical that something happened that made it stop working. The task is to find what changed, what component failed, the point where power was lost and to replace the bad part. The next step is then to reflect upon how the event may be kept from happening yet again. The actual troubleshooting process should be reviewed for potential improvements, that is, work smarter not harder.

SAFETY FIRST

Do no harm. Follow all safety procedures. The following are important and deserve consideration.

■ Do a job safety analysis (JSA).
■ Wear personal protective equipment.
■ Make sure that a fire, electrical, or explosion hazard is not created.
■ Back-up anything at risk of loss or damage in automated systems.
■ Limit the number of tools and parts entering a malfunctioning machine.

A troubleshooting plan should be developed each time a machine fails. The following suggestions will help locate the source of trouble much quicker.

■ Learn how the equipment operates.
■ Obtain and review an accurate trouble-symptom description.
■ Recreate the symptom.
■ Do the appropriate maintenance.
■ Take steps necessary to determine the root cause of the failure.
■ Repair or replace the defective component.
■ Test to confirm that the trouble has been eliminated.
■ Confirm that no new troubles have been introduced.
■ Develop a plan to prevent the future occurrence of the same problem.
■ Review one' efforts with the intent of making them safer and more productive.

WHAT TOOLS ARE NEEDED

When the word "tool" is mentioned some think of meters, hand, and power tools. Troubleshooting requires one to be highly skilled in using mental tools as well as hand and power tools. Figure 16–1 shows many of the hand tools commonly used by commercial and industrial maintenance personnel. Before starting to troubleshoot, a few of the physical and mental tools one will need are:

▦ A good complement of hand tools
▦ Appropriate test equipment
▦ Machine-specific documents and electrical diagrams
▦ Knowledge of what the equipment is supposed to do and its sequence of operation
▦ Knowledge of electrical theory and the machine's components
▦ An inquisitive mind

Tool Organization

Tool organization should consider the following: the priority of the tool, how the tool is accessed, and the locations where specific tools are needed. Tools most frequently used should be made more accessible. Some tools such as ladders and extension cords, that are used less often, may need to be stored in a location central to where they are commonly used. Each tool should be kept in the same specific location. This helps one to quickly scan the tool storage area, to determine if a tool is missing.

Arranging tools for quick and easy access saves time. The idea is to have only what is truly needed, where and when it is needed. The physical tools needed to troubleshoot are fewer than those needed to make repairs.

Skill Limits and Troubleshooting Ability

Even with the best hand tools, diagrams, and a multimeter, progress will be at a snail's pace if one lacks knowledge of troubleshooting procedures and mental skills.

It is the person, not the tool, who gets the job done. If one continues to live, and continues to work, one will need to continue to learn.

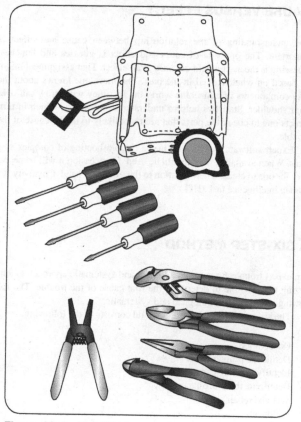

Figure 16–1   Hand Tools

CAUSE VERSUS EFFECT

The understanding of the relationship between cause and effect is important. The opposite of cause and effect is guesses and hunches. Guessing is the assignment of cause to an effect. That assignment might be based on what worked in the past. The more one knows about the way components of a machine work, the less they will throw salt over their shoulder. Statistics help to understand probability, which in turn directs one to check the parts that are most likely to be the cause of the trouble.

Expert software systems aid in the troubleshooting of complex systems. When available, they should be used. On occasion it will be necessary for one to create a new solution to the trouble at hand. Creativity is a human intelligence task (HIT).

A SIX-STEP METHOD

An expert troubleshooter uses a logical and systematic approach to analyzing the machine to determine the root cause of the trouble. The following approach is both logical and systematic.

The six-step troubleshooting method consists of the following:

▨ Observation
▨ Define the problem area
▨ Identify possible causes
▨ Determine the probable cause
▨ Test and repair
▨ Follow up

Gather information regarding the equipment and the problem. It is easier to analyze faulty operations when one knows how a machine should operate. Equipment manuals and drawings are great sources of information. Equipment repair history should also be reviewed.

Step 1—Observe

Through careful observation and a pinch of reasoning, the specific component can be identified with minimal testing. Look for visual signs of damage, chafed wires, loose components, or overheating. Smell for burnt insulation. Listen to the sound of the equipment. Check the variables such as amperage, pressure, temperature, and voltage of suspected components.

Look and See the Obvious

Thirty percent of electrical troubles can be discovered by a visual inspection. Slowly look the machine over closely. Investigate what is seen, heard, and smelled. What is there, its setting, condition, and what may be missing.

- Read and investigate error messages.
- Investigate disconnected wires, cables, missing covers, and the like.
- Look for burned, broken, disconnected, or oxidized components.
- Investigate fluid leaks.
- Listen for abnormal sounds.
- Look for physical damage that could cause a short circuit.

Pay particular attention to areas where past repairs have been made. Past problems are not necessarily the current problem, but they may provide vital clues to the current trouble. Confirm what the actual trouble is. Trouble reports are sometimes only partially correct. Make notes of any mode of operation that is not working correctly. Table16–1 is a what, when, and where guide for use at this stage in the troubleshooting process. Asking questions is very important.

Table 16–1 What, When, and Where Questions

	What is Happening	What Should Be but Is Not Happening	DIFFERENCES	CHANGES
WHAT				
WHERE				
WHEN				

Step 2—Define the Problem Area

Logic and reasoning must be joined with observations to help focus on a specific area of the machine. Often when equipment malfunctions, certain parts will work while others will not.

The key to conducting effective observations is to eliminate from further investigation, parts of the machine that are operating correctly. A copy of the electrical diagram can be used as a checklist to help focus testing not of the good parts, but only the possibly bad parts. Starting with the entire machine as the problem area, take each noted observation and ask, "What does this tell me about the area of the machine in which the trouble is located?"

When individual parts have been eliminated from the problem area, make a list of only the parts that have not been checked. Keep focusing on where the trouble is and not where it is not.

Step 3—Identify Possible Causes

Once the problem has been boxed into one area, develop a list of the parts in that area of the machine that could cause the current type of trouble. Table 16–2 is an example of a symptom, cause, and indications chart. Develop one for each of the potentially defective parts in the area under consideration.

Table 16–2 Symptom, Cause, and Indications Chart

Symptom:
Breaker tripped fuse open
Cause:
Shorted motor winding
Shorted wire
Weak breaker
Indications:
Burnt spot on wire, wire shorted to ground, shorted open, phase to phase

Step 4—Determine the Most Likely Probable Cause

Prioritize each item on the list of possible causes. The following will help in prioritizing possible causes.

Although it is possible for two parts to fail at the same time, it is not likely. Start by looking for one faulty component as the most likely cause.

▦ Look at parts that burn out or have a tendency to wear out: contacts, belts, hoses, bearings.
▦ The next most probable cause of failure are the electrical windings. They generate heat and in time, fail.
▦ Check connections for looseness, high temperature, or high resistance or discolorations.
▦ Look for defective wiring, areas where the wire insulation has been damaged.

A source of information seldom investigated is broken and worn-out parts.

Step 5—Test and Repair

Once the most probable cause is determined, one must either prove it to be, or not to be, the cause of the problem. This can be done by careful visual inspection and by the proper use of test instruments. Chapter 15, Electrical Test Instruments, can help in understanding how to use test instruments more effectively.

Before conducting any test measurement type test, one must know what the correct and incorrect answers are. Be sure the circuit is in an electrically safe work condition, before contacting or disconnecting any component or wires. Select the proper replacement part and review original part specifications. Confirm the proper setting of the new part and that it operates correctly. With repairs completed, test all functions of the machine to be sure there are no other faults. It is embarrassing to say that the problem has been repaired, only to have a new trouble suddenly pop up two days later.

Step 6—Follow Up

Determine the reason for the malfunction by asking questions such as:

- Did the component fail because of old age?
- Did the environment contribute to the failure?
- Are there wear points that caused the wiring to short out?
- What caused the part to fail and how can that cause be eliminated or reduced?
- Is there a design flaw that causes the same component to fail repeatedly?
- Did the way the machine is being operated potentially cause the failure?

By preparing, planning, and following a logical troubleshooting process, and following-up, the troubleshooter's efforts will improve.

17

ACCESS, FIRE, AND BURGLAR ALARM SYSTEMS

This chapter covers several types of electrical systems associated with facilities access control, fire detection, annunciation systems, and security systems.

FACILITIES ACCESS CONTROL

All facilities, no matter their size or industry, control the access of both authorized and unauthorized personnel to parts of an operation. While alphanumeric keypads, magnetic strips, keyless locks, and biometric systems are increasing in number, the common keyed lock is still the most economical security device in use today, but it does not offer the greatest level of security, flexibility, or the greatest number of security features. Video surveillance provides a real-time stream and historical record of who has been where and when. When coupled with a bidirectional audio system it allows security personnel to communicate with those entering and leaving a facility. Alphanumeric keypads can provide some of the benefits of a physical key as well as a record of past presence. Magnetic-stripe card systems offer some of the benefits of a keyed lock, along with the ability to be preprogrammed with a set expiration date to better control access and maintain historical records of who, when, and where. All of these systems have been successfully applied to facilities across the globe in various types of industries. Sensors of various types such as infrared, motion, magnetic, pressure, and tremble switches are routinely used.

The use of biometric data and radio frequency identification (RFID) systems represents the most advanced commercial products available today. These systems can include keychain fingerprint readers, proximity cards, and vehicle-mounted wireless transponders.

The basic functions of these types of systems are control, detection, and alarm. Operation of one or more sensors energizes or deenergizes a relay powering up a local and/or remote alarm, telephone dialer signaling security, or fire protection personnel. As these systems must remain operational during brief power outages they are provided with battery back-up power supplies. These systems can provide either local or silent alarms. Both have their strong and weak points.

FIRE ALARM SYSTEMS

The governing standard for the installation of fire alarm systems is NFPA 72, National Fire Alarm Code. Building codes determine fire alarm requirements for the various types of building and campus-like facilities. Building codes may also require activation of fire dampers, smoke dampers, and exhaust and pressurization of occupied spaces. Fire alarm systems are composed of sensors, pull stations, multiple alarms, and a central control panel. These systems are designed to detect both open and ground fault circuits. Sensors, called smoke alarms, are available in either photoelectric and ionization types. Photoelectrical sensors respond best to smoldering or slow-moving fires. Ionization-type smoke detectors respond best to flaming, fast-moving fires—the ones that develop from some other source, such as a kitchen fire or fireplace fire. These are the types of smoke detectors most commonly found in new homes. Temperature sensors are also used in dusty areas that typically result in false alarms being produced by photoelectric smoke detectors. Some smoke detectors combine both photoelectric and ionization-type sensors.

Signal Initiation

A signal may be sent to the fire alarm panel by either of three types of devices: manual, automatic, and extinguishing system activation.

Notification

When a fire is detected in a building the occupants must be properly notified. This notification can be accomplished by prerecorded and live voice (emergency voice alarm communications systems—EVACS), visible, or audible signals or some combination of these methods. Where an audible device is used it must be at least 15 dBA above the normal sound level in the space with a limit of no more than 120 dB. Where visual notification is provided by use of a strobe light, as required by the Americans with Disability Act (ADA), spacing is required to be in accordance with NFPA 72. Candela values are typically either 15, 35, 60, 75, 90, or 120 cd.

Detection

The two types of detection are automatic and manual. Red-colored manual pull stations are required to be located between 3.5 and 4.5 feet above the floor, and within 5 feet of each egress of each floor of the structure. Automatic means such as smoke detectors and heat sensors, and fire suppression (wet or dry pipe systems) activation devices are typically used. Where sprinkler systems are used, water supply valves, fire pump status, tank levels, air pressure and building temperatures are typically supervised. Supervisory signals must be either visual or audible, and must be different from both trouble and alarm conditions.

ALARM SIGNALS

An alarm signal is considered to be an emergency condition that requires immediate action. An alarm condition may be initiated by water flow, manual alarm station, or smoke or heat detectors.

Supervisory Alarm

Supervisory alarms are considered to be needed for some action to be taken relating to a fire suppression system, equipment, or maintenance of some portion of the fire detection system. A supervisory alarm may be caused by either a valve switch, tank level, near freezing building temperature, air pressure in a dry pipe system, or some type of switch connected to a valve handle.

Trouble Alarms

A trouble signal should be taken seriously as it indicates a problem with the fire alarm panel, the system's wiring, or connected devices that may render the system to be inoperable. The following events typically result in a trouble signal being produced: loss of power to the fire alarm panel, open circuit of a supervisor circuit, or low battery voltage.

SYSTEM MONITORING

Since fire alarms are considered to be directly linked to life safety and to provide a level of protection to real property, their health should be monitored. The ability to check the integrity is accomplished by end-of-line resistors, relays, and communication between the panel and the field wiring and devices.

Class A Circuits

A Class A fire alarm circuit effectively feeds power to the circuit in a loop formation. Should one wire fail the remained to the circuit is provided with power from the remaining circuit. With a loop circuit, two wires feed power to one end of the circuit while two more wires feed the same power from the opposite end of the circuit forming a loop. This method provides two paths for power to flow to all loads.

Class B Circuits

With a Class B fire alarm circuit power is supplied in a radial manner, that is, power radiates from a single point. All sensors in the circuit downstream of a break in the circuit will become inoperable. The last device in a Class B circuit has an end-of-line resistor installed between the two terminals of the device. When a break in the circuit occurs, a trouble condition will be indicated by the panel.

Smoke Detector Spacings

NFPA 72 requires that smoke detectors be placed in specific locations and at specific distances apart. Figure 17–1 provides a diagram of these allocation requirements. This national safety standard must be consulted before installing or maintaining any fire alarm system. As a very broad rule of thumb detectors should be spaced no more than 30 feet apart, no more than 12 inches from a flat ceiling, and not less than 4 inches from any corner and not less than 3 feet from any HVAC grill and more than 12 inches from any light fixture ballast. Detectors generally provide protection for areas of not more than 900 square feet. See manufacturers' installation instructions.

Notification Appliance Circuit

A notification appliance circuit (NAC) provides power to local visual and audible annunciation devices in the event of a fire. The fire alarm panel's normal power supply and battery power supply must have the ability to provide all of the loads (horns, bells, and strobes) served by the panel without excessive voltage drop (80 percent of rated voltage, typically 20.8 V).

HUMAN MACHINE INTERFACES

There are three basic methods of communication between the fire alarm panel and humans. These are light emitting diodes (LEDs), liquid crystal displays (LCDs), and a graphic layout of the building.

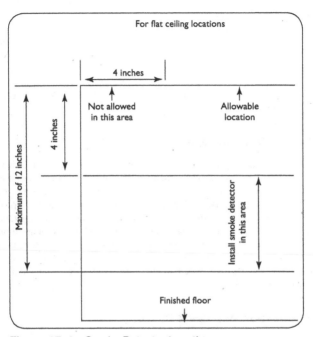

Figure 17-1 Smoke Detector Locations

These annunciation devices provide a limited amount of information as to various conditions of the fire alarm system. In general they will indicate the zone which is in an alarm, or trouble condition, silence, and reset functions. LCDs and LEDs commonly provide some amount of system programming to be accomplished from a few push buttons, or an alphanumeric keypad. Graphic layouts or floor plans of the building are many times located near the building entrance. They typically have a small lamp indicating the zone in alarm or trouble state. These graphics help fire-fighting personnel quickly determine the location of the fire.

NOTIFICATION CIRCUIT VOLTAGE DROP

When a fire alarm system is installed, a panel must be provided that is capable of providing the power demanded by the panel to all field devices. The size and length of the various conductors connecting the field devices to the panel must be selected so that the fully loaded circuit voltage drop is below 2.5 V. There are various methods of calculating voltage drop in circuits. Software programs and manual calculations are both commonly used. Conductor sizes range from 22 to 12 AWG. The goal is to provide the most distant device with a voltage well above its rated minimum operating voltage.

MASS NOTIFICATION SYSTEMS

A mass notification system (MNS) has the capability to communicate with large groups of people in or around a building or facility such as a nuclear power plant, petrochemical plant, hospital, or military base. These systems are intended to provide both prerecorded and real-time messages so as to reduce or eliminate the possibility of mass casualties. The governing standard for these types of systems was first developed by the U.S. Air force in 1989, (UFC 4-021-01). These types of systems are intended to be autonomous from the fire alarm system. The MNS system can take control of annunciation devices such as strobe lights and voice systems to communicate real-time information vital to personnel. These types of systems require that conductor integrity be monitored for various portions of a fire alarm system. MNS requirements have now been added to the 2007 edition of NFPA 72. New construction of government buildings use a fire alarm system that incorporates the functions necessary for MNS. These systems are required to be able to interface with community-wide area MNS. These systems temporarily override the fire alarm system's audible and visual signals. U.S. Army facilities use clear strobes for fire and amber strobes for MNS. The U.S. Navy uses only clear strobes. The rapid advances in communications across various platforms such as cellphones allow them to be used in emergency situations by MNSs.

CONDUCTORS FOR FIRE ALARM SYSTEMS

Article 760 of NFPA 70 and NFPA 72 have both established numerous requirements for conductors used with fire alarm systems that must be followed for these critical systems to perform as designed. Cables can be installed in various types of metallic and nonmetallic conduits. When installed outside of a conduit they must be listed as being either plenum, nonplenum, or circuit integrity (CI). CI-rated cables have been tested to perform under a two-hour fire test. Where the cable is to be installed inside of an environmental air space or plenum, it must be plenum-rated as to flame spread and the amount of smoke developed. NFPA 70 requires that abandoned fire alarm cables not identified for future use be removed from the building. This is done to help reduce the fuel load and smoke developed within the building. Fire alarm conductors are not allowed to be secured to the outside of any electrical conduit. Article 760 of the National Electrical Code specifies the following amperage limits for 18 AWG, 7 A; 16 AWG, 10 A; 14 AWG, 17 A; 12 AWG, 23 A; and 10 AWG, 28 A. Where conductor sizes 18 through 8 AWG are installed in the vertical raceways they shall be supported no more than 100 feet apart. The NEC requires that conduit systems be installed complete before any conductors are installed. The NEC also requires that fire alarm systems be installed in a neat and workman-like manner.

18

COMPUTER VOICE AND DATA WIRING SYSTEMS

INTRODUCTION

This chapter covers the wiring methods by which desktop computers are configured as standalone islands, hardwired to a network, and connected wirelessly to a network. Telephone wiring will also be reviewed. There are over 40 different types of quick connectors used to connect various types of cables to computers. The more common ones will be reviewed. Most of these connectors are "gender types" That is, they are available in both a male and a female version. In many cases, there are several types of gender adaptors and various types of adaptors available that convert one type to another.

HOT SWAPPING COMPUTER SIGNAL CABLES

Computers use voltages of around 5 V in the signal cables, and wattages in the 250 to 500 mA range. With USB and 3.5 mm connectors, no damage will result by connecting or disconnecting an energized cable, known as hot swapping. However, it is always safer to turn off the device from the keyboard then to just unplug the connector. Some systems recommend that devices such as external memory storage units (jump drives and thumb drives) be closed at the keyboard before they are removed. While disconnecting and reconnecting the cable can be done with the device in use, it is always safer to turn the device off first and then disconnect-connect the cable.

CONNECTING A DESKTOP COMPUTER AS AN ISLAND

Computer towers are typically connected to 120 AC power by use of a cord with a NEMA 5–15 male cord cap and typically an IEC C-13 female socket end, as seen in Figure 18–1. There is a family of devices known as *peripheral devices* that are connected to a computer tower. Figure 18–2 shows the various ports these peripheral devices are plugged into. The most common ones are a video monitor, keyboard, and mouse. Additional peripheral devices such as cameras, external thumb drives, headphones, IPods, microphones, MP-3 players, printers, scanners, and webcams also can be connected from time to time. Various types of cable connectors are used to connect specific types of peripheral devices to computer towers.

Video Monitors

One or more video monitors are used with desktop computers to provide a visual output of information to the operator. This information travels over electrical conductors connected to a VGA 15 pin connector seen in Figure 18–3. A cathode-ray tube ranging in size from 10 to 24 or more inches (measured diagonally) is connected to an AC wall outlet with a NEMA 5–15 plug, and by a signal cable from the monitor to the computer tower. This signal cable is provided with an analog VGA port that has 15 pins, arranged in three rows of five pins each for the red, green, and blue color channels. The VGA connector has been around since the 1980s but is being replaced by the digital video interface (DVI) connector. The most current DVI-I connector is capable of transmitting both digital-to-digital and analog-to-analog signals.

Keyboards

A keyboard is a type of human machine interface (HMI) that allows instructions to be sent to the computer tower. These instructions are

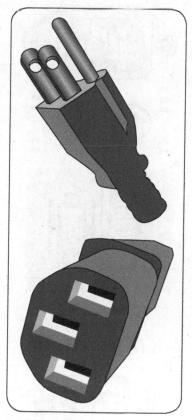

Figure 18–1 NEMA 5–15-IEC C-13 Cord Set

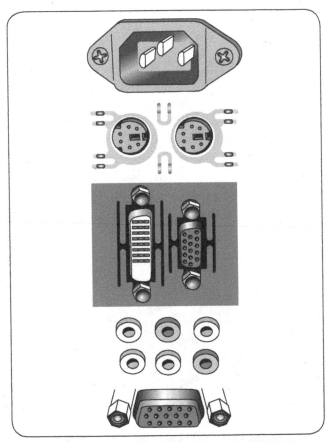

Figure 18–2 Computer Peripheral Connection Ports Diagram

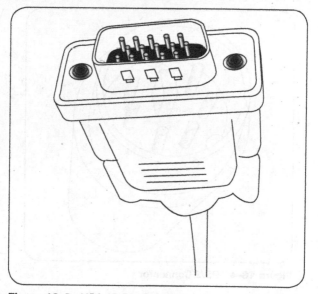

Figure 18-3 VGA 15-Pin Connector

transmitted as electrical signals that travel over conductors which are connected to the computers various parts with special ends, called connectors as seen in Figures 18–4 and 18–5. Keyboards are signaling devices and do not require 120-V power. Wired-type keyboards are provided with a cable with a purple colored PS/2 connector, or a USB connector.

Mouse

A mouse is another type of HMI that allows for the input of commands to the computer tower. This input device is connected to the computer using a small connector seen in Figure 18–6. The one on the left is

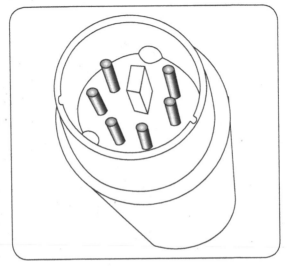

Figure 18–4 PS/2 Connector

Figure 18–5 USB Connector

Figure 18–6 A PS/2 and a USB Connector

known as a USB connector, and the one on the right is known as a PS/2 connector. PS/2 is no longer used in newer systems. A mouse controls the movement of the cursor or pointer on a display screen. Older wired mice had a green colored PS/2, while newer models are provided with a USB connection.

Printers and Scanners

To provide a method of outputting information from an electronic computer as a hard copy on paper, a printer is first connected to 120 VAC source of electrical power. Most of the time a small step-down transformer, equipped with a NEMA 5–15 male connector as seen in Figure 18–7, and a USB-type B mini-female connector from the transformer as seen in Figure 18–8, connects to the printer. Many times the connection of the printer to the tower is made by use of a coax 12-V connector. The printer or scanner is connected to the tower by use of a cable with a USB-type B mini-connector at the printer/scanner and a standard USB-type A connector on the end that connects to the computer tower. The Original USB 1.01 standard provided data transfer rates of up to 12 Mbit. The USB 2.0 provides data transfer rates of up to 480 Mbit and the newest USB 3.0 will provide data transfer rates of up to 4.8 Gbit. Both 2.0 and 3.0 are compatible with previous versions.

External Audio Speakers

Towers also may be provided with a means of exporting (outputting) an audio signal. The connector seen in Figure 18–9 is typical of ones used

Figure 18–7 Typical Printer Transformer

for audio output. Speakers may draw power from a small transformer, or from the tower. Today speakers are typically connected to the tower with a USB connector. Earlier models of speakers were connected to the tower with a green-colored 3.5 mm TRS connector using a USB connector. Power is typically limited to about 2.5 W.

Figure 18–8 Coax 12-VDC Connector

Figure 18–9   Type-A and Mini-Type-B USB Connectors

Microphones

Today there are various types of application (software) that use a microphone to input commands to the computer tower, it is connected to the tower using a connector as is seen in Figure 18–11. One of the more common is Dragon NaturallySpeaking. This application allows a person to speak to the computer using a microphone to input words

Figure 18–10   3.55 mm TRS Connector Used with Speakers

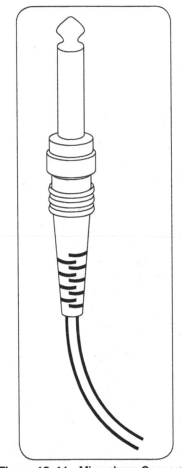

Figure 18–11 Microphone Connector

into an MS Word document. This eliminates much of the need to type text. Microphones are typically provided with a pink 3.5 mm TRS connector.

Webcams and Microscopes

On occasion an electronic microscope may be required to view objects up close under magnification. A common electronic microscope is seen in

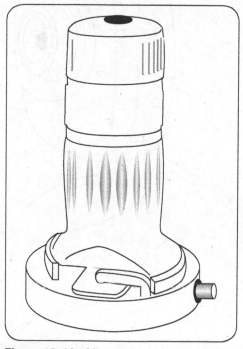

Figure 18–12   Microscope

Figure 18–12. With the development of Internet-based video conferencing, web cameras can be found connected to desktop computers by means of a USB connector. A common web camera is seen in Figure 18–13.

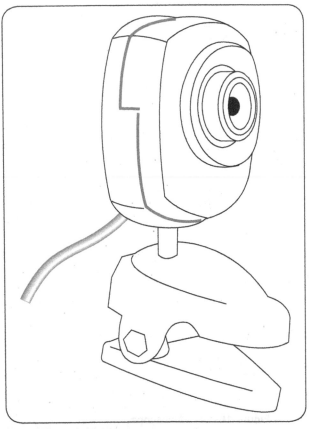

Figure 18–13 Webcam

CONNECTING A DESKTOP COMPUTER TO A WIRED NETWORK

The following is a discussion of additional steps that need to be taken once a desktop has been cabled together as an island. Connection of a desktop computer to a wired network requires that a network interface card (NIC), illustrated in Figure 18–14, be installed in the tower cabinet. The network cable is connected to the wall jack and is connected to the tower with an RJ-45 Ethernet connector seen in Figure 18–15. The letters R J stand for registered jack, which has eight positions and eight conductors. Ethernet cables are classified as being Cat-5 (Category) 10/100 Base T cable, which supports data traffic of up to 100 ms per second. Cables can be wires straight through, that is, pin number 1 on one end is connected to pin number 1 on the other end. This can be confirmed by holding both ends of the cable in hand with the clips down and looking to see if the same color of wire is in the same position on both ends. If they are not then the cable is connected as a cross-connect or crossover pattern. Review the documentation for the desktop to determine the correct cable to be used. Crossover cables cannot be used to connect a computer to a wired network.

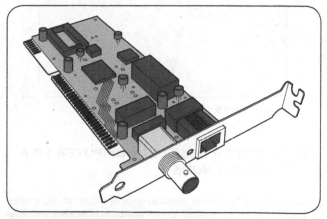

Figure 18–14 Network Interface Card

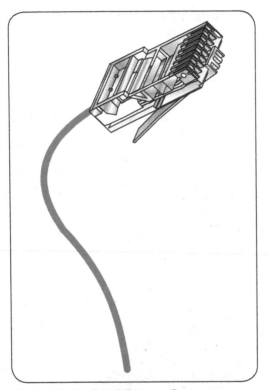

Figure 18–15   RJ-45 Ethernet Connector

CONNECTING A DESKTOP COMPUTER TO A WIRELESS NETWORK

Using radiowaves, a wireless router, as seen in Figure 18–16, transfers packets of data between the wired network and one or more wireless desktop/

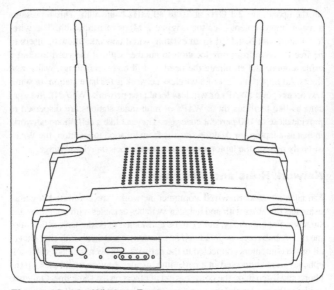

Figure 18-16 Wireless Router

laptop computers. Typically a router has a small external transformer that is connected to a 120-VAC wall outlet, and then to the router. A cable is run from one computer connected to the wired network to the router. The cable connecting the wired network to the router uses an eight-pin RJ-45 Ethernet connector. The Ethernet cable uses four pairs of unshielded twisted stranded conductors covered by an outer plastic jacket. While gray is the most common color for the outer jacket, blue and red are also used.

The router's software must be loaded on a desktop computer for the router to begin to transfer data to and from any computer equipped with a wireless card. In 1997, the Institute of Electrical and Electronics Engineers developed the rules by which wireless communications are handled in its 802.11 standard. Over time various updates to this standard have been 802.11a, b, g, with the latest being 802.11n. Data transmission frequencies for the update are 2.4 GHz at 20 to 40 MHz bandwidth. This represents a major improvement over the original 2 Mbps of bandwidth. The wireless feature can be added to an existing wired network, allowing users to be free to move from one location to another without the need to drag a cable around, and eliminate the need to drill holes in ceilings, walls, and floors. An important part of a wireless network is the base station, or wireless access point (WAP) or wireless local area network (WLAN). In larger campus-like facilities these WAPs or mini-radio stations are dispersed to provide close to 100 percent coverage. They act like a cell phone network, in that as a laptop (or desktop) moves from location to location, the WAPs instantly pick up the laptop that comes within their individual range.

Network Hubs and Switches

Virtually all modern wired computer networks are connected together using lots of cables with and hubs, or switches or cables with both hubs and switches. For simplicity, hubs can be considered to be the least efficient of the two. A hub receives data from one computer and then sends that data to all of the computers connected to the network. This is much like when telephones were connected in a party-line manner. Anyone could pick up the phone and listen in on the conversation of others on the same line. Only two parties—the sender and the receiver—can talk to each other. If more people start talking, the data gets jammed up. A switch is much smarter than a hub. A switch looks at each data packet (like snail-mail letter) and directs the packet only to the port which the destination computer is connected to on the switch. A switch provides better use of the available bandwidth. Switches allow for two-way communication to occur at the same time (full duplex). Hubs allow communication in one direction at a time (half duplex) and are much slower devices. Hubs are less costly than switches. Routers are a more advanced type of network device and are used to connect computers on one network to computers on another network.

TYPES OF COMPUTER NETWORKS

The two types of computer networks encountered in commercial and industrial facilities are local area networks (LANs) and wide area networks (WANs). LANs (HUB) are built to service several computers in a small area, such as a home, a single building, or several buildings. LANs (switch) use Ethernet technology and are most commonly owned by a nontelephone company. WANs cover much larger distances and are many times connected to telephone communications networks. WANs use technology such as ATM, Frame Relay, and X.25 to cover larger distances. The Internet is a WAN. LANs can be connected together. LANs can be connected to WANs, and WANs can be connected to the public Internet. Figures 18–17 and 18–18 provide diagrams for both a hub and switches.

TELEPHONE VOICE COMMUNICATIONS SYSTEMS WIRING

The modular quick connect connector used with modern telephone wiring is the RJ-11. The most common cable connected to an RJ-11 connector is an untwisted flat satin two-pair (four wires) 28 AWG stranded

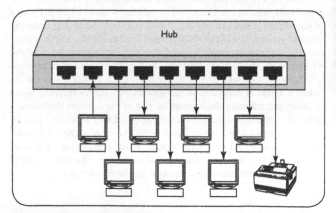

Figure 18–17   Network Hub

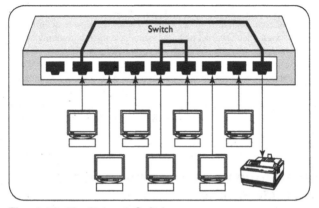

Figure 18–18   Network Switch

conductor cable rated for up to 150 V. The RJ-11 as seen in Figure 18–19 is a six-position connector that uses only four or two of the six positions. Typically only two of the four wires are used, and only four of the connector's six positions are used. As is the case with all of the RJ series of connectors, actual pin-out and colors used vary depending upon the standard applied. Use of the red and green insulated wires, and the center two pins, is the more common application. Pin-out diagrams for all of the RJ connectors are always viewed with the hock clip down and the pins up.

When working with any of the modular quick connectors in the RJ family it almost is necessary to use specialized tools for stripping wires in cables and crimping the connectors. Use of the common electrician's "Klein's" or lineman's pliers is not recommended for conductors in the 16 to 28 and smaller AWG range. There is less chance of making a mistake when using an RJ-45 or RJ-11 crimper. When using twisted pairs, do not untwist a pair for more than one-half of an inch. The more a pair is untwisted, the more interfering crosstalk will be encountered.

REFERENCE MATERIAL

This chapter contains electrical reference materials. The first portion of the chapter contains electrical formulas with worked examples for many of them. They are presented in an alphabetical order of the desired variable. For example, the first items are those where the value desired is amperage (I). The remainder of the chapter provides completed time-saving charts.

AMPERAGE-CURRENT

$I = \dfrac{\sqrt{P}}{R}$: Where $\sqrt{\ }$ = the square root, I = current in amperes,

P = power in watts, R = resistance in ohms

Example: $I = ?$ P = 800 watts, R = 300 ohms: $I = \dfrac{\sqrt{800}}{300} = \dfrac{2.66}{}$ = 1.63 amps

$I = P/E$: Where P = power in watts, E = voltage

Example: P = 400 watts, E = 120 volts: I = P/E 400/120 = 3.33 amps

$I = E/R$: Where E = Voltage, R = resistance in ohms

Example: E = 120 volts, R = 6 ohms: I = 120 ÷ 6 = 20 amps

$I^2 = \dfrac{P}{R}$: Where I^2 = the value squared, P = power in watts,
R = resistance in ohms

Example: P = 300 watts, R = 80 ohms: I^2 = 300/80 = 3.75 amps

$I = \dfrac{\sqrt{AP}}{E}$: Where $\sqrt{}$ = square root, AP = apparent power in volt-amperes, E = voltage

Example: AP = $\sqrt{720/240}$ = 3 amps

$I = E/Z$: Where E = voltage, Z = impedance

Example: E = 240 volts, Z = 7.1 ohms: I = 240/7.1 = 33.8 amps

$I = \dfrac{\sqrt{AP}}{Z}$: Where AP = apparent power in volt-amps, Z = impedance

Example: I = $\dfrac{\sqrt{720}}{7.11}$ = 101.4/ 7.1 = 10 amps

$I = VA/E$ **Single phase:** Where VA = volt-amps, E = volts

Example: 4500 VA, 240 volts: I = 4500/240 = 18.75 amps

$I = Va/E \times 1.732$: Where Va = Volt amperes, 1.732 = three-phase factor

Example: 7200 Va, 208 E: I = 7200/360 = 200 amps

$I^2 = P/R$: Where I^2 = amperage squared, P = power in watts, R = resistance in ohms

Example: P = 300 watts, R = 80 ohms: I^2 = 300/80 = 3.75 = I^2 $\sqrt{3.75}$ = 1.936 amps

$I = VA/E \times 1.732$: Where VA = apparent power in volt-amps, E = voltage, 1.732 = constant for three-phase

Example: VA = 500, E = 120: I = 500/120 = 2.4 amps

$I = VA/E \times 1.732$: Where VA = volt-amperes, E = voltage, 1.732 = constant for three-phase

Example: VA = 72,000, E = 208: I 72,000/208 = 200 amps

Table 19-1 Amperes per Phase for 1 kW at Various Voltages and Power Factors for Single-Phase System

Volts	Power Factor					
Voltage	P f 1.0	.90	.8	.7	.6	.5
120	8.55	9.25	10.41	11.90	13.88	16.66
208	4.81	5.34	6.01	6.87	8.02	9.62
220	4.54	5.05	5.68	6.49	7.58	9.09
230	4.35	4.83	5.44	6.21	7.24	8.70
440	2.27	2.52	3.84	3.24	3.79	4.54.
460	2.18	2.42	2.72	2.72	3.62	4.35

Table 19-2 Amperes per Phase for 1 kW at Various Voltages and Power Factors for Three-Phase System

Volts	PF 1	.9	.8	.7	.6	.5
208	2.78	3.08	3.47	3.97	4.63	5.56
220	2.63	2.92	3.47	3.97	4.63	5.56
230	2.51	2.79	3.14	3.59	4.18	5.26
440	1.314	1.460	1.640	1.877	2.189	2.628
460	1.26	1.39	1.57	1.79	2.09	2.51

Carbon Brushes Used in Motors and Generators

Guideline for pressure: 1.5 lb per square inch of brush area. Total PSI + brush contact area should equal 1.5 PIS/square inch.

CAPACITANCE FORMULAS

1 Farad = 1,000,000 micro-farads (μF); 1 micro-farad = 1/1,000,000 of a Farad

Capacitors in Parallel, Three or More

$$CT = C\text{-}1 + C\text{-}2 + C\text{-}3 \ldots$$

Example: C-1 = 8, C-2 = 5, C-3 = 11: CT = 8 + 5 + 11 = 24, CT = 24

Capacitors, Two Connected in Series

CT = C-1 × C-2/C-1 + C-2: CT = capacitance total, C-1 = capacitance of the first capacitor, C-2= capacitance of the second capacitor,

Example: C-1 = 20, C-2 = 8: CT = 20 × 8/20 + 8 = 160/28 = CT = 5.71 μF (micro-farad)

Capacitors, Three or More Connected in Series

$$CT = 1/C\text{-}1 + 1/C\text{-}2 + 1/C\text{-}3 \ldots$$

Three or more: C-1 = 20, C-2 = 8, C-3 = 5: CT = 1/20 + 1/8 + 1/5

$$= 0.5 + 0.125 + 0.2 = 825\ \mu F$$

Capacitors Connected in a Combination of Series in Parallel

Step 1. Calculate capacitance of the parallel branch: CT = C-1 + C-2 + C-3 + …
Step 2. Calculate the capacitance of the series combination:

$$CT = C\text{-}1 \times C\text{-}2/C\text{-}1 + C\text{-}2$$

CONDUCTOR CROSS-SECTIONAL AREA IN CIRCULAR MILS (CMIL)

$$CMIL = 2 \times K \times I \times L / VD$$

Where CMIL = circular mils, K = constant 12.9 ohms for copper, I = amps, L = length of wire one way, VD = voltage drop

CMIL = K × I × D/VD: Where K = 12.9, I = 50 amps, D = 200 circuit feet (total length), VD = 4.4

Example: CMIL = 12.9 C × 50 × 200 /4.4 = 29,318. # 6 AWG = 26,250, # 4 AWG = 41,742. Select larger, not smaller wire CMIL.

Conductor Resistance in Ohms

R = 2 × L × 12.9/CMIL: Where R = resistance in ohms, 2 – constant, L = wire length, 12.9 = constant. CMIL = conductor cross-sectional area in square inches

Conductor Cross-Sectional Area in Square Inches

A = D² × .7854. Where A = area in square inches, D² = diameter squared, .7854 = constant

Example: Diameter of # 10 AWG = 0.116 sq in: A = 0.116 × 0.116 = 0.01356 × .7854 = diameter of conductor = 0.11 sq in. See NEC, Chapter 9, Table 9 for conductor circular mil areas.

DIRECT CURRENT FORMULAS

I = Hp × 746/ V × Eff: Where I = current in amps, Hp = horsepower, 746 = watts per horsepower, V = volts, Eff = efficiency as a decimal

Example: Hp= 5, V = 220, Eff = .87: I = 5 × 746/ 220 × .87 = 3730/191.4 = 19.5 amps

kW = I × V/1000: Where kW = kilowatts, I = current, V = voltage, 1000 = constant

Example: 20 amps, 240 volts: kW = 20 × 240/1000 = 4.8 kW

I = KVA × 1000/V: Where kVA = kilo-volt-amps, V = voltage

Example: 15 kVA, 240 volts: I = 15 × 1000/240 = 62.5 amps

I = P/V: Where I = current in amps, P = power in watts, V = voltage

Example: 1500 watts, V = 120: I = 150/120 = 1.25 amps

Horsepower = (V × I × Eff)/746: Where V = volts, I = current, Eff = efficiency, 746 = watts per Hp

Example: 120 V, 3 amps, .85 Eff: Hp = 120 × 3 × .85/746 = 3 × 6/746 = .41 Hp

Hp out = I × V × Eff: Where I = current in amps, V = volt, Eff = efficiency as a decimal

Example: 120 volts, 3 amps, .85 Eff: Hp = 3 × 240 × .85/746 = 3060/746 = 4.1 Hp

P = E × I: Where P = power in watts, I = current, E = volts

Example: 120 volts, 17 amps: P = 120 × 17 = 2040 watts

Eff = (746 × HP)/V × A: Where Eff = motor efficiency as a %, 746 = watts per Hp, Hp = motor horsepower

Example: 5 Hp, 240 volts, 20 amps: Eff = (746 × 5)/ 240 × 20 = 3730/4800 = .777 = .78 Eff

Resistors total (RT) in series: R-1 + R-2 + R-3 . . . Where R-1 is the resistance of the first resistor, R-2 is the resistance of the second resistor, R-3 is the resistance of third resistor.

Example: R-1 = 5 ohms, R-2 = 8 ohms, R-3 = 11 ohms: RT = 5 + 8 + 11 = RT = 24 ohms

Resistor total (RT) in Parallel:
$$RT = 1/ \ 1/R\text{-}1 + 1/R\text{-}2 + 1/R\text{-}3$$

Where R-1 = resistance of the first resistor, R-2 = resistance of the second resistor, R-3 = resistance of the third resistor

Example: RT = R-1 = 12 ohms, R-2 = 10 ohms, R-3 = 8 ohms

RT = 1/12 + 1/10 + 1/8, RT = 0.083 + 0.01 + 0.125 = RT = .218 ohms
Power = $I^2 \times R$: Where I^2 = current squared, R = resistance in ohms

Example: I = 7 amps, R =19: P = $(7)^2 \times 19$, P = 49 × 19 = 931 watts

Electrical Values

kV:	=	kilovolts = 1,000 volts:
kW:	=	kilowatts = 1,000 watts
1 mA:	=	1 milli-amp = 1/1000 of one amp
kΩ:	=	one thousand ohms of resistance
μF:	=	micro-farads = 1 one-millionth of 1 farad
Mhz:	=	megahertz, one million Hertz

Ω = ohms, electrical unit of resistance to pass one ampere, at a pressure of one volt.

A = ampere, unit of quantity of electrons flowing in circuit, also denoted as I

P = power in units of watts, or kilowatts

Pf = power factor, expresses relationship of real power to reactive (apparent) power

E = volt in AC systems, sometimes V is also used, V should be used only with DC circuits.

Table 19-3 Equipment Grounding Conductor Size

Overcurrent Protective Device Size in Amps	AWG
15	14
20	12
30–60	10
70–100	8
110–200	6
227—300	4

FUSE OPENING TIME

As a general rule of thumb as to how fast a fuse will blow when overloaded can be determined from Table 19-4.

Table 19-4 Fuse Opening Time

Fuse Rating	At 135% Overload	At 200% Overload
0–30	60 seconds	2 seconds
31–60	60 seconds	4 seconds
61–100	120 seconds	6 seconds
101–200	120 seconds	8 seconds
201–400	120 seconds	10 seconds
401–600	120 seconds	12 seconds

Generator Frequency (Hz) GHz = number of poles × RPM/120;
Where RPM = speed of the motor, 120 = constant

HEAT PRODUCED IN A CONDUCTOR PER WATT

Calorimeter tests show that when one watt of electrical energy flows for one minute, 0.057 BTUs will be generated by the current flow in a conductor.

IMPEDANCE FORMULAS

The impedance (Z) in an AC circuit is the result of inductive and capacitive reactance, eddy currents, skin effect, and resistance of the conductors in the circuit.

$Z = AP/I^2$: Where Z = impedance in ohms, AP = apparent power in VA, I = current in amps

Example: AP = 1800 VA, I = 7.5 amps: Z = 1800/7.5 × 7.5 = 1800/56.25 = 32 ohms

$Z = E^2/AP$: Where Z = impedance in ohms, E = volts squared, AP = apparent power in VA

Example: E = 240, AP = 1800 VA, Z = 240 × 240 = 57600/1800 Z = 32 ohms

$Z = E/I$: Where Z = impendence in ohms, I = current in amps

Example: E = 240 volts, I = 7.5 amps: Z = 240/7.5 Z = 32 ohms

$Z = \sqrt{R^2 \times X_L^2} \times X_L^2$: Where Z = impedance in ohms, R^2 = resistance squared, X_L = inductive reactance in ohms squared

Example: R = 40 ohms, X_L = 30 ohms: $Z = \sqrt{40^2 + 30^2} = 1600 + 900 = 2500$, $Z = \sqrt{2500} = 50$ ohms

$\mathbf{Z = (R^2 + (X_C - X_L)^2)}$: Where Z = impedance in ohms, X_C = capacitive reactance, X_L = inductive reactance

Example: 240 volt motor, resistance of 80 ohms, X_L = 90 ohms, 30 ohms: Z = 80² + (30 – 90)² = (80 × 80) – (30 – 90 = 60)² 6400 – 3600 Z = 2800 ohms

$$\mathbf{Z = \sqrt{R^2 \times XL^2}}$$: Where R² = resistance, squared in ohms, X_L^2 = inductive reactance squared

Example: R = 0.13 ohms, 1/0 copper wire in aluminum conduit, X_L = 0.44 ohms

$$Z = \sqrt{(13)^2 \times (0.44)^2} = 169 \times 0.1936, Z = \sqrt{169 \times .1936} = 32.72 = Z$$
$$\sqrt{32.73\,Z} = 5.72 \text{ ohms}$$

$$\mathbf{Z = \sqrt{R^2} + X_L^2}$$: Where R = resistance in ohms, X_L = inductive reactance

Example: R = 0.50 ohms, X_L = 60 ohms: Z = $\sqrt{0.25 + 36}$ = 36.25 ohms, Z $\sqrt{36.35\,Z}$ = 6.02 ohms

IMPEDANCE OF A CONDUCTOR IN A CONDUIT

Chapter 9, Table 9 of the NEC lists Z per 1000 feet of wire. Z = R/1000 × Distance (length) for PVC, aluminum, and steel conduits. Caution: Table 9 has values for both copper and aluminum; impedance increases from PVC, to Aluminum, to Steel conduit.
Example: Z # 12 AWG in PVC conduit, Distance = 800 ft = 2/1000 ft: Z = 2/1000 = 0.002 × 800 Z = 16 ohms

KW TO AMPS

$I = kW \times 1000 \times Pf /E$: Where Pf = power factor as a decimal, E = volts, I = amps, kW = one thousand watts, $100 = 1k$

Example: $kW = 30$, $E = 240$ volts, $PF = .85$: $I = 30 \times 1000 \times .85/ 240 = 30 \times 1000 = 30,000$, $30,000 \times .85 = 25,500$, $25,500 \div 240 = 106.25$ amps

KVA TO AMPS

$I = kVA \times 1000 / E$: Where I = current in amps, kVA = one thousand volt-amperes, 1000 = constant, E = voltage

Example: $kVA = 30$, $E = 240$: $I = 30 \times 1000 = 30,000$, $30,000 \div 240 =$, $I = 125$ amps

MOTOR FORMULAS FOR THREE-PHASE AC MOTORS

$kVA = I \times E \times 1.732/1000$: Where I = current in amps, E = voltage, 1.732 = constant for three-phase motor, 1000 = constant

Example: $I = 6$ amps, $E = 220$ volts: $kVA = 6 \times 220 \times 1.732/1000 = 2286/1000 = kVA = 2.3$

$kW = I \times E \times 1.732 \times PF/1000$: Where kW = thousands of watts, E = voltage, 1.732 = three-phase constant, PF = power factor as a decimal, $1,000$ = kilo factor

Example: $I = 8$ amps, $E = 220$ volts, $PF = .85$: $kW = 8 \times 220 \times 1.732 \times .85/1000 = 2591/1000$

kW = 2.6

$I = kW \times 1000/1.732 \times E$: Where E = volts, 1.732 = three-phase constant, kVA = thousand volt amperes

Example: 15 kVA, E = 208 volts: $I = 15 \times 1000/1.732 \times 208 = 15,000/360.3$
I = 41.6 amps

$I = Hp \times 746$ watts per Hp/$1.732 \times E \times Eff \times Pf$: Where Hp = motor nameplate horsepower, 1.732 = three-phase constant, E = volts, Eff = efficiency of the motor as a decimal, Pf = power factor of the motor

Example: Hp = 10, V = 208, Eff = 85, Pf = .90: $I = 10 \times 746/1.732 \times 208 \times .85 \times .90 = 7460/275.6 = 27$ amps

$Hp = I \times E \times Eff \times Pf/746$: Where HP = horsepower, I = current, Eff = efficiency of the motor, PF = power factor, 746 = watts per horsepower

Example: I = 8 amps, E = 240, .85 Eff = .89 Pf: $Hp = 8 \times 240 \times .85 \times .90 / 746 = 1453/746 = 1.95$ Hp

$kW = I \times E \times Pf/1000$: Where kW = thousands of watts, I = current, Pf = power factor, 1000 = constant

Example: I = 17 A, E = 240 V, .85 Pf: $kW = 17 \times 240 \times .85/1000 = 3.47$ kW

$I = kW \times 1000/E \times Pf$: Where kW = kilowatts, 1,000 = kilo constant, Pf = power factor

Example: 8 kW, 240 V, Pf .85: $8 \times 1000/240 \times .85 = 8000/204 = 39.22$ amps

$FLA = Hp \times 746/E \times Eff \times Pf$ for a single-phase motor:
Where Hp = motor nameplate horsepower, 746 = watts per horsepower, E = voltage, Eff. = efficiency of the motor, Pf = power factor

Example: HP = 5, E = 460, Eff = .89, Pf = .85: $5 \times 746/460 \times .89 \times .85 = 3730/348 = 10.72$ amps

MOTOR FORMULAS

$$Hp = E\text{-}3 \times I\text{-}3 \times Eff \times Pf/ 746$$
$$E\text{-}3 = Hp \times 746/ I\text{-}3 \times Pf \times Eff$$
$$I\,3 = HP \times 746 / E3 \times Pf \times Eff$$
$$E = Hp \times 746/ Pf \times I\text{-}3 \times E\text{-}3$$
$$E = \text{phase-to-phase voltage}$$
$$I = Hp \times 746/1.732 \times E \times Eff \times Pf$$
$$I = kW \times 1000/1.732 \times E \times Pf$$
$$I = kVA \times 1000/ 1.732 \times E$$
$$kW = I \times E \times 1.732 \times Pf/1000$$
$$kVA = I \times E \times 1.732/1000$$
$$Hp\ out = HP \times I \times E \times 1.732 \times Eff \times Pf /746$$

Motor Formulas for Single-Phase Motors

$$I = Hp \times 746/E \times Eff \times Pf$$
$$I = KW \times 1000/E \times Pf$$
$$KW = I \times E \times Pf/1000$$
$$KVA = I \times E/1000$$
$$HP = I \times E \times Eff \times Pf/746$$
$$\textbf{Horsepower} = \textbf{(E} \times \textbf{I} \times \textbf{Eff} \times \textbf{Pf)/746}$$

Example: 20 amps, 240 volts, .89 Eff: Hp = 20 × 240 × .85/746 5.5 Hp

Efficiency of a motor as a % = Output power/ Input power

Example: 1240 watts output, 1680 watts input: Eff = 1240/1680 = .74%

Motor Starting and Running Currents

$$I = E/R \text{ starting}$$

Example: Starting current I = 240/0.5 = 480 amps

Example: Running R = 0.5 ohms, V = 240 volts : $Z = \sqrt{R^2 + 6^2} = \sqrt{0.25 + 36} = 6$ ohms, I = 240/6 = 40 amps

$$I = E/Z: \text{ Where } Z = \sqrt{(R^2 + X_L^2)} \text{ running current}$$

I = current in amps, E = volts, R = resistance in ohms, Z = impedance in ohms, XL = inductive reactance in ohms

Example: Running R = 0.5 ohms, E = 240 volts: $Z = \sqrt{R^2 + 6^2} = \sqrt{0.25 + 36} = 6$ ohms, I = 240/6 = 40 amps

HP = (E × I × Eff)/ 746: Where Hp = horsepower, E = volts, I = current, Eff = Efficiency, 746 = watts per horsepower

Example: V = 240 volts, I = 12 amps, Eff = .90: 240 × 12 × .9/ 746 = 3.5 Hp

Motor Amps, Three-Phase

I = (Hp × 746)/ 1.732 × E × Eff × Pf: Where I = current, HP = horsepower, 746 = watts per horsepower, 1.732 = constant, Eff = efficiency, Pf = power factor

Example: Hp = 50, V = 440, Eff = .86, Pf = .9: I = 50 × 1.732/ 440 × .86 × .9 = 63.24 amps

Motor-Locked Rotor Current, Three-Phase, True Power

LRA = 1000 × Hp × kVA/HP/ E × 1.732 × PF × Eff
Where LRA = locked rotor amps, 1000 = constant, HP = horsepower, kVA = thousand volt-amps, E = volts, 1.732 = constant, Pf = power factor, Eff = efficiency of the motor

Motor-Locked Rotor Current, Three-Phase Apparent Power

LRA = 1000 × Hp × kVA × Hp/ E × 1.732

Where kVA = kilo-volt-amps, E = voltage, Hp = motor nameplate horsepower, 1.732 = three-phase power constant

Motor Horsepower, Three-Phase

Hp = 1.732 × I × E × Eff × Pf/746: Where Hp =horsepower, 1.732 = three-phase constant, I = current, E = volts, Eff = efficiency, Pf = power factor, 746 = watts per horsepower

Motor Horsepower, Single Phase

HP = E × I × Eff/746: Where Hp = horsepower, E = volts, I = current, Eff = efficiency, 746 = watts per horsepower

Example: E = 240, I = 15 amps, Eff = .85

$$Hp = 240 \times 15 \times .85/746 = 3060/746 = 4.1 \text{ Hp}$$

Motor Horsepower Where RPM and Torque are Known

HP = RPM × T/ 5252: Where 5252 = constant, RPM = speed of motor, T = torque of the motor

Example: RPM = 1750, T = 37 in pounds (lbs): Hp = 1750 × 37/5252 = 64750/5252 = 12.3 Hp

Motor Torque

T = Hp × 5252/RPM: Where 5252 = constant, T = torque, Hp = horsepower

Example: Hp = 60, RPM = 1725: T = 60 × 5252/1725 = 315,120/1725 = 228.35 in pound feet (lb ft)

$$T = I \times RPM/5252$$

Motor-Starting Torque

Ts = HP × 5252/RPM × %: Where % = starting torque as a % of full load torque, Hp = motor nameplate horsepower, 5252 = constant, RPM = speed of the motor, Ts = torque of the motor starting

Example: Hp = 5, RPM = 1750, Slip = 5%: Ts = 5 × 5252/1750 = 26,260/1750 = 15.0 × 250 – 37.5 lb ft

Motor Nominal Torque Rating

T = Torque in pound inches (lb in): HP × 63,000 /RPM: Where 63,000 = constant, Hp = motor nameplate horsepower, RPM = speed of the motor

Example: Hp = 5, RPM = 1750: T = 5 × 63,000/1750 = 315,000/1750 = Torque 180 lb in

T in lb ft = HP × 5252/RPM: Where T = torque in pound feet, Hp = horsepower, 5252 = constant, RPM = revolutions per minute

Example: T = 10 × 5252/1725 = 52520/1725 = T = 30.4 lb ft

Motor Horsepower

Hp = RPM × T/5252: Where Hp = horsepower, RPM = revolutions per minute, T = torque, 5252 = constant

Example: 60 Hp, 1725 RPM, Torque = 15.2: Hp = 1725 × 15.2/ 5252 = 4.992 or 5 Hp

Table 19-5 Motor Full-Load Current for Three-Phase Motors

HP	208 Volts	230 Volts	460 Volts
1	4	3.6	1.8
1.5	5.7	5.2	2.6

Table 19-5 (Continued)

HP	208 Volts	230 Volts	460 Volts
3	10.6	9.6	4.8
5	16.7	15.2	7.6
7.5	24	22	11
10	31	28	14
15	46	42	21
20	59	54	27
30	88	80	40
40	114	104	52
50	143	130	65
60	169	154	77
75	211	192	96
100	273	248	124

Table 19-6 Motor Full-Load Current for Three-Phase Motors

Characteristic	+ 10% of rated	−10% of rated
Starting current	+ 10%	−10%
Full load amps	7%	+11%
Torque	+20%	−20%
Effiency	Minor	Minor
RPM	+1%	−1.5%
Temperature rise	−3 °C	− 6 °C

Motor Synchronous Speed

RPM = Hz × 120/Np: Where RPM = revolutions per minute, Hz = frequency, 120 = constant , and Np = number of poles in the stator winding

Example: Hz = 60, Poles = 4

RPM = 60 × 120/4 = 7200/4 = 1800 RPM. Actual will be less due to slip.

Motor RPM Speed, Less Slip

RPM = 120 × Hz × (100 – slip)/NP: Where RPM = revolutions per minute, 120 = constant, Hz = frequency, Np = number of poles in the stator winding

Example: Hz = 60, Np= 4, Slip = 5%

120 × 60 × (100 – 5%)/4 = 120 × 60 = 7200 × .95 = 6840/4 = 1710 RPM

Motor Slip Speed

Slip = Synchronous speed – actual speed/ synchronous speed

Example; Slip = 1800 –1700 /1800 = 100/1800 = 5.5 % Slip

Table 19-7 Motor Acceleration Time

Motor Frame Size	Maximum Time (in Seconds) without Overheating
48–56	8
143–286	10
324–326	12
364–505	15

Table 19-8 Motor Synchronous Speed at 60 Hz

Number of Poles	RPM
2	3600
4	1800
6	1200
8	900
10	720
14	514
18	400
20	360
24	300

Table 19-9 Motor Voltage Variation Impact upon Starting Current and Starting Torque

Voltage % of Rated	% of Nominal Starting Current	% of Starting Torque
40	112	32
60	250	72
80	450	128
100	700	200

Table 19-10 Motor Starting Torque of NEMA Design Letter

NEMA Design Letter	% of Full-Load Torque
A	165
B	150
C	225
D	275

Horsepower (Hp) of Three-Phase Motors

HP = (1.732 × E × I × Eff × Pf/ 746: Where Hp = horsepower, 1.732 = three-phase constant, E = volts, I = current, Eff = efficiency, Pf = power factor, 746 = watts per horsepower

Hp = Motor nameplate horsepower, 1.732 = three-phase constant, E = volts, I = amps, Eff = efficiency of the motor, Pf = power factor of motor, 746 = watts per horsepower

Example: E = 240, I =12 amps, Eff = .85, Pf = .9

HP = 1.732 × 240 × 12 × .85 × .9/746 = 38159/746 = 5.1 Hp

Motor Current

I = Hp × 746/ E × Eff × Pf: Where I = motor current, Hp = horsepower, 746 = watts per horsepower, eff = efficiency, Pf = power factor

Example: 10 Hp, 240 v, Eff =.9%, Pf = .88

I = 10 × 746/240 × .90 × .88 = 39 amps

Motor Operating Cost

Cost/Hr = Watts × Cost/ kWH/1000

Where watts = watts of power consumed per hour, Cost = cost per 100 watts or kHW, 100 = constant to remove kilo.

Example: Cost per kWH = 16 cents, power consumed per hour in watts = 5,000

C/H = 5,000 × 16/1000 = Ch = 80 cents hour

Motor-Driven Load Speed

Driver RPM/ Driven RPM = Driver pitch diameter/Driven pitch diameter

S1 = Driver RPM S2 = Driven RPM

P1 = Driver pitch diameter, P2 = Driven pitch diameter

S1/S2 = P2/P1

Driver RPM

To determine the driver RPM, multiply the driving pitch diameter by the RPM of the driven, then divide by the driving pitch diameter.

Driven RPM

To determine the driven shaft RPM, multiply the driving pitch diameter by the RPM of the driver and divide by the RPM of the driven pitch diameter.

Driven Pitch Diameter

To determine the driven pitch diameter, multiply the driven pitch diameter by the RPM of the driven shaft, then divide by the RPM of the driving shaft.

Power Formulas

P in watts = E/I: Where P in = power input to the motor in watts, E = voltage, I = current

Example: E = 240, I = 7: P = 240 × 7 = Power = 1680 watts

Pa = I × E: Where Pa = power apparent, I = current, E = volts

Example: I = 18 amps, E = 240 volts: Pa = 18 × 240 = 4320 (VA)

Pa = E²/R: Where Pa = power apparent, E = volts, R = resistance in ohms

Example: $(240)^2 = 57,600/13.3 = Pa = 4331$ VA

Pa three-phase = EL × Il: Where Pa = power apparent, E = volts phase, Il = current in the line, where EL is the voltage of the line.

Power apparent = E phase × I phase × 1.732

Example: El = 240, Il = 7 amps: AP = 240 × 7 = 1680 VA

Three-phase AP = 240 × 7 × 1.732 × .97 = 2531 VA

Power Loss

Power loss = P out − P in: Where P out = power output, P in = power input to the motor

Example: Watts in = 500, watts out = 418, = 82 watts power lost

P true = I^2 × R: Where P true = true power, I = current in amps, R = resistance in ohms

Example: $(18)^2 × 13.3 = 324 × 13.3 = Pa\ 4256$ VA

P = I^2 × R: Where P = power, I = current, I = current, R = resistance

P = E^2 /R: Where P = power, E = voltage, R = resistance in ohms

Table 19-11 Power Factor Data

Power Factor	Amps	% Current Increase	Wire Size	Energy Loss Increase
1	100	0	100	-0-
.9	111	11	+23%	20%

Table 19-11 *(continued)*

Power Factor	Amps	% Current Increase	Wire Size	Energy Loss Increase
.8	125	25	+56%	56%
.7	143	43	+204 %	104%
.6	167	67	+279 %	179%
.5	200	100	+400%	300%

From Table 19-11 it can be understood that the power factor has a significant impact upon the energy lost due to heating, and that as this heating is increased, the current goes up, and the wire size must be increased.

Power Factor

$$Pf = \text{True power/Apparent power} = \text{Watts/Volt amperes}$$

Example: Watts = 60, VA = 80: Pf = 60/80 = Pf = .75

Power Loss Due to Voltage Drop

P loss = I × Ed: Where I = current in amperes, Ed = voltage drop for both wires

Example: I = 14 amps, Ed = 1.83: P loss = 14 × (1.83 + 1.83), P = 14 × 3.16 Ed, 50.5 Watts

Cost of Power Due to Voltage Drop

P loss = I^2 × R: Where I^2 = current squared, R = resistance in ohms

Example: I = 20 amps, R = 0.3 ohms: p = (20 × 20) × 0.3 = 400 × 0.3 = 120/1000 = 0.12 kW

Cost 8.6 cents per 100 watts = 0.086: 0.086 × 0.12 = 0.032 per Hr: 0.01023 × 24 = 0.24768 × 365 = $ 90.40 per year

Power True in Watts, Single Phase

$Pt = E \times I \times Pf$: Where Pt = true power, E = volts, I = current, Pf = power factor

Example: E = 24 volts, R = 2 ohms: $P = E \times I$: $Pt = E \times I \times Pf$; $24 \times 2 \times .8$ Pt = 384 watts

Reactance Capacitive X_C

$X_C = 1,000,000/ 2 \times 3.1416 \times Hz\ 60 \times C$: Where X_C = capacitive reactance in ohms
$1,000,000$ = constant, 2 = constant, 3.1416 = constant, C = capacitance in micro-farads (μF)

Example: F = 60, C = 13 μF: $X_C = 1,000,000/2 \times 3.1416 \times 60 \times 13 = 204$ ohms

$$X_C = 1/ 2 \times 3.1416 \times Hz \times C$$

Example: F = 60 Hz, C = .000022 farads = 22 μf: $X_c = 1/ 2 \times 3.1416 \times 60 \times .000022 = X_c$ =72 ohms

Reactance Inductive X_L

$X_L = 2 \times 3.1416 \times Hz \times L$: where X_L = inductive reactance in henry

Example: Hz = 60 Hz, H = 10

$$X_L = 2 \times 3.1416 \times 60 \times 10 = 6.28 \times 10\ X_L = 62.8 \text{ ohms}$$

Resistance DC and AC

Generally the DC resistance and the AC resistance, for conductors 1/0 and smaller, are essentially the same.

Resistance

$R = E^2/P$: Where R = resistance in ohms, E^2 is the voltage squared, P = the power in watts

Example: P = 870 watts, E = 240 volts: $R = 240^2 = 57,600/870 = 66.2$ ohms

$R = P/I^2$: Where R = resistance in ohms, P = watts, I^2 is the current squared

Example: P = 870, I = 3.6 amps: R = 870/12.96 = R = 67 ohms

$R = E/I$: Where R = resistance in ohms, E = voltage, I = current

Example: E = 24 volts, I = 2: $R = E \times I$: 24/2 = 12 ohms

Sine Wave Values

Effective volts = .707 × PV
Peak volts   = 1.414 × Effective volts
Effective volts = Average × 1.11
Average volts = .9 × Effective volts
Average   = .637 × Pv
Peak volts   = 1.57 × Average volts

Temperature Conversion

F to C = C = (F – 32)/ 1.8
C to f = (1.8 × C) + 32

TRANSFORMER FORMULAS

Transformer Voltage Regulation

Percent voltage regulation is the drop in voltage from no load to full load as a percentage. Voltage regulation is typically between 2 and 4 percent.

$$\% \ Er = (Enl - EFL) \ / \ EnL \times 100$$

Example: EnL = 240, EfL = 220: Er = (240 – 220) = 9 × 100 = 9%

A voltage drop of some 9 percent may have a negative impact upon equipment that is sensitive to voltage changes.

Transformer Delta Three-Phase Primary Current

Ip = kVA × 1000/ 1.732 × Ep: Where Ip = primary current, kVA = transformer kVA rating, 1000 = kilo constant, 1.732 = three-phase constant

Example: kVA = 500, three-phase, Ep = 480: Ip = 500 × 1000/1.732 × 480 = 500,000/830 , Ip= 603 amps

Is = kVA × 1000/ 1.732 × Es: Where kVA = Transformer kVA rating, 1000 = kilo constant, 1.732 = three-phase constant, Es = voltage of the secondary

Example: 500 kVS, Es = 208: Is = 500 × 1000/1.732 × 208 = 500,000/360, Is = 138.8 amps

Is = Ip Ep/Es: Where Is = current of the secondary winding, Ip = current of the primary, Ep = voltage of the primary, Es = voltage of the secondary winding

Example: Ip = 104 amps, Ep = 480 volts, Es = 208 volts

Is = 104 480/240 = 104 × 2 = 208 Is = 208 amps
Ip = kVA × 1000/ Ep: Where kVA = Transformer kilo-volt-amps, Ep = voltage of the primary, 1000 = kilo constant, Ep = voltage of the primary

Example: kVA = 50, Ep = 480: Ip = 50 × 1000/480. Ip = 104 amps

Transformer Delta versus Wye

Delta-phase voltage is the same as the line voltage. E phase = E line.

Wye-phase current is the same as the line current. Line voltage is greater than phase voltage by a factor of 1.732.

$$E\ line = E\ phase \times 1.732\ I\ phase$$

Transformer Secondary Phase Currents, Three-Phase

I phase = kVA/ E line × 1.732: Where Ip = phase current, kVA = thousand volt-amps of the transformer, E line = voltage of the line, 1.732 = constant

Example; 150 kVA, 208, three-phase: I phase = 150,000 VA/ 208 × 1.732. = 208 × 1.732 = 360.26

$$150,000/ 360.26 = 416\ amps.\ I\ phase = 416\ amps$$

Transformer Primary Line Current

IL = VA/(E × 1.732): Where IL = line current, VA = transformer VA rating, E = voltage, 1.732 = constant

Example: 1 kVA, E = 480, three-phase: I line = 100,000/ 480 × 1.732 = 100,000/831.36 = I line = 120 amps

Transformer Line Current Delta-Connected Three-Phase 120/240

Ip = VA/E: Where Ip = primary current, VA = transformer volt-ampere rating, E = voltage

Example: 100 VA, E = 240: I phase = 50,000/24-0 = I phase = 208 amps

Transformer Phase Current Delta-Connected

I phase = VA phase/E phase: Where VA = phase of the transformer VA rating, E phase = voltage of the phase

Example: 100 kVA, three-phase: VA phase = 100,000/ 3 = VA phase 33,333, I phase = 33, 333/480 = I phase = 69.4 amps

Transformer Secondary Neutral Current

Unbalance load, delta L-2 is high leg

$$In = IL\text{-}1 - I\ line\ 3: \text{Unbalanced wye: } In = \sqrt{\ IL\text{-}1^2 + IL\text{-}2^2 + I\ L\text{-}3^2\ -}$$
$$[C\ Il\text{-}1 \times IL\text{-}2) + (Il\text{-}2 \times IL\text{-}3) + (IL\text{-}1 \times IL\text{-}30)]$$

Example: What is the secondary line current of a delta-connected transformer? kVA = 100, 480/240 voltage three-phase

I line = VA/ (E × 1.732), I line = 100 kVA/ 240 × 1.732 I line = 240 amps

Transformer Delta and Wye Voltages

Delta: Line and phase voltages are the same.

Wye: Line voltage is greater than phase voltage by a factor of 1.732. Line voltage is 208 with phase voltage of only 120.

Transformer Delta-Wye Currents, Three-Phase

Delta: Line current is greater than phase current by a factor of 1.732. This is due to two phase windings feeding current to each phase.

Wye: Line and phase winding current are the same.

Transformer Primary Current

Ip = Ns/ Np × Is: Where Ip = current of the primary, Np = number of turns of the primary, Ns = number of turns of the secondary

Example: Ns = 88, Np = 1760, Is = 5: Ip = 88/1760 = 0.05 × 5 = .25 amps

Transformer Formulas

$$Es/Ep = Ns/Np$$
$$Ip/Is = Ns/Np$$

Where Np = number of turns of the primary, Ns = number of turns of the secondary, Ep = voltage of the primary, Es = voltage of the secondary, Ip = current of the primary, Is = current of the secondary

Primary voltage = ratio × secondary voltage
Ratio = primary tap setting/secondary rating

Example: Ratio primary tap setting = 2400, secondary voltage rating = 240: ratio 2400/240 = ratio = 10:1

Pf = W/VA: Where Pf = power factor, W = watts, Va = volt-amperes
I = W/ E × Pf: Where I = current, E = voltage, Pf = power factor
E = W/I × Pf: Where E = voltage, W = watts, I = current, Pf = power factor
VA = W/Pf: Where VA = volt-amperes, W = watts, Pf = power factor

Transformer Delta Three-Phase

IL = Ip × 1.732: Where IL = line current, Ip = current in the primary, 1.732 = constant
Ip = IL/1.732: Where Ip = primary current, IL = current in the line, 1.732 = constant

Transformer Wye Three-Phase

Ep = EL/1.732: Where Ep = volts of the primary, El = volts of the line, 1.732 = constant
EL = Ep × 1.732: Where EL = volts of the line, Ep = volts of the primary, 1.732 = constant

Transformer Turns Ratio

Delta-Delta TTR = 2:1, Primary 480 volts, Secondary 240 volts
Delta-Wye TTR = 4:1, Primary 480 volts, Secondary 120 volts

Transformer Secondary Amps Single- and Three-Phase

Secondary amps, single-phase (Is) = VA/E
Secondary amps, three-phase = VA/Volts × 1.732 × Z. Where VA = volt-amperes of the transformer, 1.732 = three-phase constant, Z = impedance of the transformer

Transformers Secondary Available Fault Current Single Phase

Is = VA/E × Z: Where VA = volt-amperes, Z = impedance of the transformer

Available Fault Current on the Secondary of a Three-Phase Transformer

VA/E × 1.732 x Z: Where VA = volt-amperes, E = volts, 1.732 = three-phase constant, Z = impedance of the transformer as a percent
I = VA/E for single-phase
I = VA/ E × 1.732 = I FC three-phase

Transformer Secondary Short Circuit Current—Isscc

Step 1. Calculate the secondary full-load current.

kVA/ Es = FLA = Is = Current of the secondary .
Step 2. 100/Z = Isscc, where Z = impedance of the transformer as a percent

Example: kVA = 150,00 Es = 240

Is = 150,000/240 = 625, Is = 625 amps

2. 100/4 = 25 × 625 = 15,625 amps, Isscc = 15,625 amps

Where motors are present in the circuit under short circuit conditions a motor acts as a source and will contribute to the short circuit current—an amount equal to about 5 times the motor's full load current (FLA or RLA).

Transformers Delta High Leg to Neutral Voltage

EhL = EnL × 1.732: Where EhL = voltage of the high leg, EnL = voltage of the neutral, 1.732 = constant

Example: EnL = 115 volts, Eh = 115 × 1.732, Eh =199.8 volts

Transformer Delta

$$\text{Phase amps (Ip)} \times 1.732 = \text{IL}$$
$$\text{Ip} = \text{Line amps (IL)}/1.732$$
$$\text{Phase volts (Ep)} = \text{El} = \text{line volts}$$

Transformer Delta, Three-Phase

Ip = IL/1.732: Where Ip = current of the primary, IL = current of the line, 1.732 = three-phase constant

IL = Ip × 1.732: Where IL = current of the line, Ip = current of the primary, 1.732 = three-phase constant

Example: Il – 4.33 amps, Ip = 4.33/1.732 = 2.5 amps

$$\text{Ip} = 2.5, \text{IL} = 2.5 \times 1.732 = 4.33 \text{ amps}$$

Transformer, Wye-Connected

$$\text{Phase amps (Ip)} = \text{Line amps (IL)}$$
$$\text{Phase volts (Ep)} \times 1.732 - \text{Line Volts (EL)}$$

Ep = EL/ 1.732: Where Ep = voltage of the primary, El = voltage of the line

Example: Ep = 120. EL = Ep × 1.732. 120 × 1.732 = EL = 207.84 volts

Is = Ep × Ip/Es, Where Is = current on the secondary winding, Ep = voltage on the primary winding, Ip = current on the primary winding

Example: Ep = 120 volts, Ip = 2 amps, Es = 12 amps

$$Is = 120 \times 2/ 12 = 120 \times 2 = 240/ 12 = Is = 20 \text{ amps}$$

Es = Ep × Ip/Is: Where Es = voltage on the secondary winding, Ep = voltage on the primary winding, Ip = current on the primary winding

Example: Ip = 1 amp, Ep = 120 volts, Is = 5 amps

$$Es = 120 \times 1/5 = 120/5 = Es = 24 \text{ volts}$$

Ip = Es × Is/Ep: Where Ip = current on the primary winding, Es = voltage on the secondary winding, Ep = voltage on the primary winding

Example: Ep = 240 volts, Es = 120 volts Is = 20 amps

$$Ip = Es \times Is/ Ep = 120 \times 20/240 = Ip = 10 \text{ amps}$$

Ep = Es × Is/Ip: Where Ep = voltage on the primary winding, Es = voltage on the secondary, Is = current on the secondary winding, Ip = current on the primary

Ip = Es × Is/Ep: Where Ip = current on the primary, Es = voltage on the secondary, Is = current on the secondary, Ep = voltage on the primary

Es = Ep × Ip/Es: Where Es = voltage on the secondary winding, Ep = voltage on the primary winding, Ip = current on the primary winding

Pf = W/VA: Where Pf = power factor, W = power in watts, VA = volt-ampere rating of the transformer

Example: Hp = 3, amps = 34, voltage = 120

3×746 = watts output, $34 \times 115 - 3910$ VA input 2238 W/ 3910 VA = Pf = .57

Ep = Es × Is/ IP: Where Ep = voltage of the primary, Es = voltage of the secondary, Ip = current of the primary winding

Example: Ip = .5, Is = 5, Es = 12

$$EP = 12 \times 5/.5 \ 12 \times 5 = 60/.6 \ Ep = 120 \text{ volts}$$

Es = Ns/Np × Ep: Where Es = voltage on the secondary, Ns = number of turns on the secondary, Np = number of turns on the primary, Ep = voltage of the primary

Example: Ns = 88, Np = 1760, Ep = 2400

$$Es = 88/1760 = 0.05 \times 2400. \ Es = 120 \text{ volts}$$

Es = Ep × Ns/ Np: Where Es = voltage of the primary, Ns = number of turns on the secondary, Np = number of turns on the primary

Example: Ns = 300, Np = 3000, Ep = 2400

$$Es = 2400 \times 300/3000 = .10 \times 2400 \ Es = 240 \text{ volts}$$

Transformers, Three-Phase Wye

EL = Ep × 1.732: Where E line = voltage of the line, Ep = voltage of the primary, 1.732 = three-phase constant
     IL = Ip: Where IL = line current, Ip = phase current

Transformers, Three-Phase Delta

IL = Ip × 1.732: Where IL = line current, Ip = phase current, 1.732 = three-phase constant
     EL = Ep: Where EL = line voltage, Vp = phase voltage

Transformer-Rated Secondary Current, Three-Phase

Is = kVA × 1000/Es × 1.732. Where Is = current on the secondary winding of the transformer, kVA = Kilo-volt-amperes of the transformer, 1.732 = three-phase constant

Example: kVA = 500, Es = 208 three-phase

Is = 500 × 1000/208 × 1.732 = 50,000/360.3. Is = 1388 amps

Transformer kVA

kVA = line-to-line voltage × I phase × 1.732/1000

Example: Line-to-line voltage = 208, Ip = 762 amps

kVA = 208 × 762 × 1.732 /1000 = 274 kVA

Transformer Secondary Current, Three-Phase

Is = kVA/ Es × 1.732: Where Is = current on the secondary winding, kVA = transformer kilo-volt-amps, Es = voltage of the secondary winding, 1.732 = three-phase constant.

Example: kVA = 15, Es = 208 volts

Is = 15,000/208 × 1.732. Is = 125 amps

Transformer Secondary Current, Single-Phase

Is = kVA × 1000/ Es: Where Is = current on the secondary winding, kVA = transformer kilo-volt-amps rating, Es = voltage of the secondary winding

Example: kVA = 25, Es = 240

Is = 25 × 1000/240 = 25,000/240, Is = 104 amps

Transformer Primary Current

IP = kVA × 1000/ Ep: Where Ip = current on the primary winding, kVA = kilo-volt-amps, Ep = voltage of the primary winding

Example: kVA = 25, Ep = 480 volts

$$Ip = 25 \times 1000/480. \quad Ip = 52 \text{ amps}$$

Transformer Temperature Class 220, 115 °C Rise

This type of transformer is designed for only a 115 °C temperature rise. A Class 220, 80 °C rise type of transformer is designed for only an 80 °C rise. This is made possible by use of a larger conductor with lower resistances. It is more effective with higher first cost.

Transformer Temperature Class 220, Class H

This allows for only a 40 °C maximum ambient, 150 °C temperature rise, plus 30 °C hot spot, which equals a maximum of 220 °C hot spot temperature inside of the transformer. This information is of little value in the field, as the actual temperature inside of the transformer cannot be measured with the transformer online. Only the temperature rise of the air across the transformer can be measured safely. So the limiting factor in the field is the temperature rise of the cooling air into and out of the transformer.

Transformer Temperature Class 185, Class F

This allows for a 40 °C maximum ambient, 115 °C rise, 30 °C hot spot, which equals a 185 °C maximum temperature deep inside of the transformer winding. This type of transformer is typically best in the range of 3 to 30 kVA, 480-volt primary, and 240/120 secondary. When motors and transformers are rebuilt today, a class F insulation is typically used do to high, short-rise time events from switch mode power supplies.

Transformer Temperature Class 150

This is a 40 °C maximum ambient, plus 80 °C rise, plus 30 °C maximum hot spot temperature. Maximum operating temperature is 150 °C.

Best for up to 2 kVA, single-phase, 480 volts primary, 240/120 secondary volts.

Transformer Temperature Rating

Maximum ambient temperature is 40 °C. 55 °C average rise of winding in the transformer. 10 °C additional for hot spot temperature in the transformer; this is the point of greatest possible insulation damage as a result of heat. This allows for a 40 °C + 55 °C + 10 °C = 105 °C, for a Class 105, or Class A transformer. Best if limited to 150 VA single-phase.

Transformer Impedance

As the impedance of a transformer increases, the amount of voltage drop from no load to full load will increase.

Variables in an Electrical System

The primary variables in electrical systems are: power, volts, amps and resistance. As the number of variables in a system increase, its complexity also increases. Secondary or additional variables in electrical systems are: power factor, efficiency, impedance, capacitive reactance, inductive reactance, single- or three-phase, transformer turns ratio, and the configuration of the transformer windings such as delta and wye. If you have some difficulty understanding electrical systems, it is not surprising as it has a dozen variables.

Volt-Amperes Three- and Single-Phase

$$VA = I \times E \times 1.732 \text{ for three-phase}$$
$$VA = I \times E \text{ for single-phase}$$

Voltage Values in Three-Phase Calculations

In many electrical formulas there is a requirement to multiply the voltage times 1.732. Table 19-12 provides the results that can be used to simplify a formula. For example, $1.732 \times 240 = 416$. By using 416, a step can be simplified.

Table 19-12 Voltage Values in Three-Phase Calculations

For	1.732	208	Use 360
	1.732	220	Use 381
	1.732	230	Use 398
	1.732	240	Use 416
	1.732	277	Use 480
	1.732	440	Use 581
	1.732	460	Use 797

$E = VA/I$. Where E = volts, VA = volt-amperes, I = current

Example: 42,000 VA, 350 amps

$$V = 42,000/ 350 = 120 \text{ volts}$$

Volt-Amperes Formulas

$VA = I \times E$. Where VA = volt-amperes (apparent power), I = current, E = voltage of the circuit

Example: 100 amps, 240 volts

$$VA = 100 \times 240 = 24,000 \text{ VA}$$

Voltage Amplitude of Sine Wave Conversions

Vp is the voltage to the peak of a signal. As an AC signal peaks in both the positive and negatives polarities, the term peak to peak is used, or Vp-p.

$$\text{Vp to VRMs: VRMs} = VP \times .707$$

VRMs to VP: VP = VRMs × 1.414

Vp to Vav, where Vav is the average of the voltage peak signal value:

Vp	= VP × 0.637
VRMs to Vp: Vp-p	= VRMs × 2.828
Vp-p to VRMs: VRMs	= Vp-p × 0.35235
Vav to VRMs: VRMs	= VA × 1.11
Vp-p to Vp: VP	= Vp-p/2
VRMs to Vav: Vav	= Vp-p × 0.318
Vav to Vp-p: Vp-p	= Va × 3.14

Voltage

E = I × R: Where E = voltage, I = current in amperes, R = resistance in ohms

Example: I = 2 amps, R = 12 ohms. E = 2 × 12 = 24 volts

E = P/I: Where E = voltage, P = power in watts (true) or volt-amperes (apparent power), I = current in amperes

Example: I = 2 amps, P = 870 watts, I = 3.6 amps. E = 870/ 3.6 = 241.7 volts

E = $\sqrt{P \times R}$: Where E = voltage, P = power in watts, or VA, and R = resistance in ohms

Example: P = 870 watts, R = 66.2 ohms. E = $\sqrt{870 \times 66.2}$. E = 239.98 volts = 240 volts

E = I × R: Where E = voltage, I = current in amperes, r = resistance in ohms

Example: I = 18 amps, R = 32 ohms. E = 18 × 32 = 576 volts

E = I × Z: Where E = voltage, I = current in amps, Z = impedance in ohms

E = AP/I: Where E = voltage, AP = apparent power in VA, I = current in amps

Example: VA = 750, I = 6 amps. E = AP/I = 750/6 = 125 volts

E = $\sqrt{AP \times Z}$: Where E = voltage, AP = apparent power in VA, Z = impedance in ohms.

Example: VA = 1250 VA, Z = 8. E = $\sqrt{1250 \times 8}$ 1250 × 8 = $\sqrt{10000}$. E = 100 volts

Voltage Imbalance

Imbalance = largest voltage difference from the average voltage. 2 percent imbalance = 10 percent increase in motor losses. 5 percent imbalance = 30 percent increase in motor losses.

VU = VD/VA × 100: Where VU = voltage imbalance, VD = largest voltage deviation in volts, VA = average voltage

Example: VD = 6 volts, VL-1 = 460, VL-2 = 454, VL-3 = 448. VD = 1363 total/3 = average voltage

L-1 to L-2 = 460 volts, L-2 to L-3 = 454 volts, L-1 to L-3 = 460 − 454 = 6 volts, 454 − 448 = 6 volts. 6/ 454 × 100 = 0.013%, in the field, this would be a very low value

Example: 445 volts, 455 volts, 460 volts, average = 453

453 − 445/ 454 = 453 − 445 = 8/453 = 1.89 % voltage imbalance

Voltage Drop (ED) Single-Phase Circuit

ED = 2 × K × L × I/CMIL: Where ED = voltage drop in volts, @ = constant, K =constant for copper = 12.9, L = conductor length, I = current in amps, CMIL = conductor area in circuit LAR mils.

Example: 240 V, 40 A, 60 ft long, 16, 510 CMIL, 12 ohms

Ed = 2 × 12 × 40 × 60/16,510. ED = 3.5 volts drop

Example: K = 12.9, I =24 amps, D = 150 feet, CMIL = 10,380 for # 10 AWG

ED = 2 × 12.9 × 24 × 150/10380. ED = 8.95volts

Voltage Drop Conductor Sizing Three-Phase

CMIL = $\sqrt{3 \times K \times I \times D}$ / **allowable ED**: Where CMIL = conductor area in circular mils, 3 is for three conductors in the three-phase circuit, K 12.9 constant for copper, I = current in amps, D = conductor length in feet.

Example: I = 18 amps, Distance = 390 ft, E = 480 volts, 3% = 14.40 volt drop allowable

CMIL = 1.732 × 12.9 × 18 × 390/ 14.4
CMIL = 10,892 = # 8 AWG per NEC Chapter 9, Table 9

Voltage Drop-Limiting Conductor Length to Minimize Voltage Drop in a Three-Phase Circuit

D = (CMIL × ED)/1.732 × K × I: Where D = conductor length in feet, CMIL = conductor circular mils, K = constant for copper 12.9, I = current in the circuit

Example: CMIL = 10,382, I = 18 amps, ED = 3% = 0.03

D = 10,382 × 0.03 = 31146, 1.732 × 12.9 = 22.3 × 18 = 402.2 = 77.5 Ft

Voltage Drop in a Three-Phase System

ED = 2 × R × L × I/ CMIL × .866: Where 2 = constant, R = resistance in ohms, L = length in feet, I = current, CMIL= conductor area circular mils

2 = constant, R = conductor resistance in ohms, L = length of the wire, I = current, .866 = constant

Example: R = 10 ohms, L = length in feet, amps = 110

2 × 10 = 20 × 75 = 1500 × 110 = 16,500/ CMIL 66,360 = 2.486 × .866 = 2.75 volt drop

Conductor Circular Mil Area, Single-Phase Circuit

CMIL = 2 × K × L × I/ED: Where 2 = constant, K = constant for copper, 12.9, L = conductor length, I = current in amps, CMIL= conductor area in circular mils

Example: ED = 3%, = 0.03 I = 18 amps, L = 120 ft

CMIL = 2 × 12.9 × 18 × 120/3 = 55,728/0.03 = 18,576 CMIL
# 6 AWG = 26,250, # 8 AWG = 16, 768. Select larger conductor size

Conductor Size in Circular Mils, Three-Phase

CMILS = 1.732 × K × L × I /ED: Where CMIL = conductor sixe in circular mils, 1.732 = three-phase factor, L = length of the conductor, I = current in amperes, ED = voltage drop, K = constant for copper 12.9.

Example: L = 250 ft., I = 7 amps, ED = 3%

CMIL = 1.732 × 12.9 × 250 × 7 / 0.03. 1.732 × 12.9 = 22.34. 22.34 × 250 = 55857
55857 × 7 = 39090. 39090/ 0.03 = 13, 033 CMILS

Voltage Drop Where Resistance, Length, and Current Are Known

ED = 2 × R × L × I/CMIL: Where 2 is a constant, R = resistance of the circuit in ohms, I = current in amps

Example: E = 208, L = 40 ft, I = 60 amps, CMIL = 16,510

$$ED = 2 \times 12.9 \times 40 \times 60/16510. \quad ED = 3.5 \text{ volts}$$

Three-phase ED: E = 208, I = 110 amps, L = 75 ft, R = 12 ohms, CMIL = 66,360, .866 = constant

ED = 1.732 × 12.9 × 75 × 110 / 66,360. = 1843281/ 66,360 = 2.8 × .866 = 2.4 volt drop.

Voltage Drop Conductor Sizing to Prevent Excessive Voltage Drop

CMILS = 2 × K × I × D/ ED: Where 2 = constant, K = constant for copper 12.9, I = current in amps, D = conductor length, ED = voltage drop

Example: K = 12.9, I =26 amps, D = 90 ft

CMIL = 2 × 12.9 × 26 × 90/ 7.2 ED = 8385 = # 10 AWG. For motor application 26 × 1.25% = 32 amps.

Voltage Drop

ED = I × R. In Chapter 9 of the NEC, Table 9 lists conductor resistances in ohms per 1,000 ft. As voltage drop is a function of the total conductor length, the constant 2 is used.

Example: # 10 AWG, 24 amps, 160 ft

R	=	(1.2)/1000 × 320 = 0.384 ohms
ED	=	I × R. ED = 24 × .384 = 9.22 volts
240 – 9.22	=	230.8 volts is applied to the load.

Three-Phase VA Given Current and Volts

VA = I × E × 1.732: Where VA = volt-amperes, I = current, E = voltage, 1.732 = three-phase constant

Example: 4 amps, 120 Volts.

$$VA = 4 \times 120 \times 1.732 = 831.4 \text{ VA}$$

Voltage Drop Where CMIL Is Known

$$ED = KIL/CMIL$$

Voltage Drop in Existing Circuit

Single phase: ED = 2 × K × I × D/ED: Where ED = voltage drop, 2 = constant, K = constant for copper = 12.9, D = distance in feet
Three-phase: ED = 1.732 × K × I × D/ED: Where ED = voltage drop, 1.732 = three-phase constant, K = 12.9 for copper, I = current, D = distance in feet

Example: K = 12.9 for copper, I = 50 amps, D = 55 ft CMIL = 41,740

$$ED = 2 \times 12.9 \times 50 \;/\; 41740 \quad ED = 2.79 \text{ existing}$$
$$7.22 - 2.79 = 4.41 \text{ maximum for the addition. } CMIL = 2 \times 12.9 \times 50 \times 55/4.41 \quad CMIL = 19{,}014 \quad CMIL = \#\,6 \text{ AWG}$$

Watts to Amps

$$A = W/E: \text{ Where A = amps, W = watts, E = voltage}$$
$$I = P/E: \text{ Where I = amps, P = power in watts, or VA, E = voltage}$$

Example: W = 1800, E = 240 volts

$$I = 1800/240 = 7.5 \text{ amps}$$

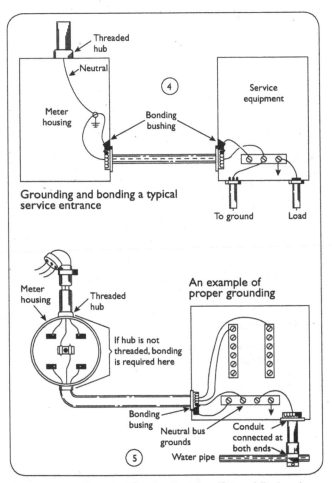

Figure 19-1 Electrical Service Bonding (Ground Rod and Ground Rod Wire Not Shown)

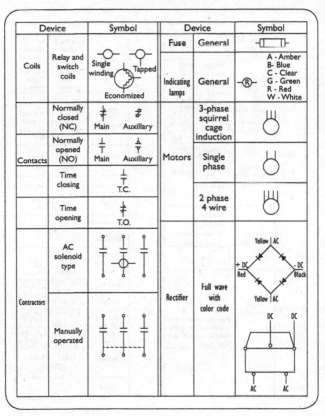

Figure 19-2 Electrical Symbols Coils, Contacts, Contactors, Fuses, Motors, and Rectifiers

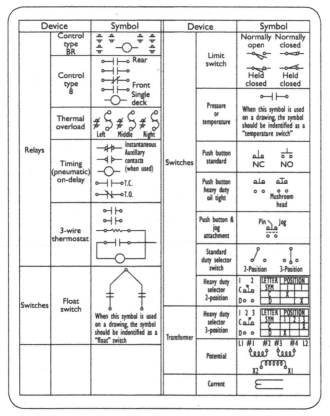

Figure 19-3 Relays, Switches, and Transformers

REQUIREMENTS OF THE DRIVEN LOAD

- Necessary horsepower
- Torque requirements
- Frequency of starts and stops
- Speed
- Mounting position—horizontal or vertical
- Direction of rotation
- Ambient temperature
- Environmental considerations

Table 19-13 Electrical Motor Requirements Considerations

Electrical Requirements
Horsepower
Speed
Voltage
Phases
Frequency
Mounting position
Ambient temperature
Environmental considerations

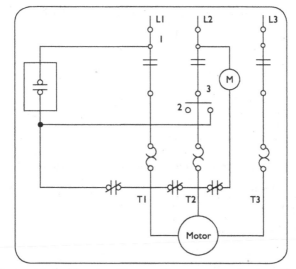

Figure 19-4 Electrical Diagram Motor Starter with Start Push Button

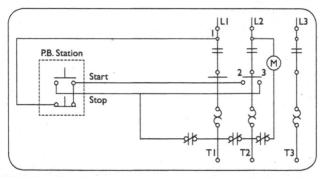

Figure 19-5 Electrical Diagram Motor Starter with Start-Stop Push Buttons

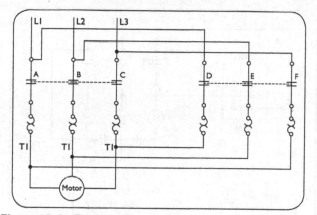

Figure 19-6 Electrical Diagram Reversing Motor with Hand-Operated Starter

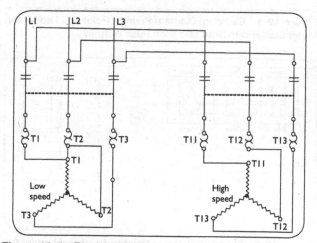

Figure 19-7 Electrical Diagram Two-Speed Motor with Low-and High-Speed Windings

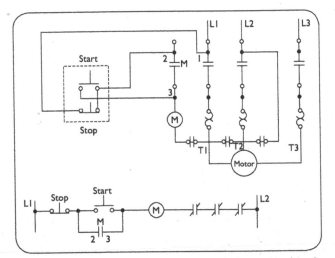

Figure 19-8 Electrical Diagram Point to Point and Ladder for Motor Starter with Start and Stop Push Buttons

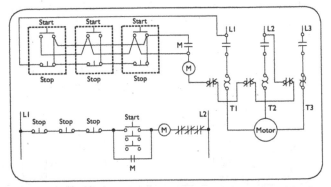

Figure 19-9 Electrical Diagram Point to Point and Ladder for Motor Starter with Multiple Start and Stop Push Buttons

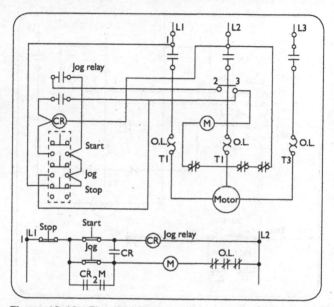

Figure 19-10 Electrical Diagram Point to Point Ladder for Jog Relay

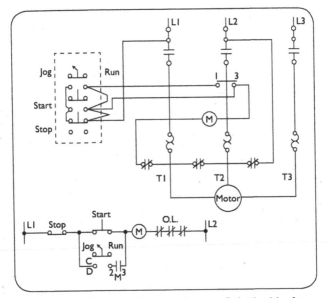

Figure 19-11 Electrical Diagram Point to Point Ladder for Jog and Run Relay

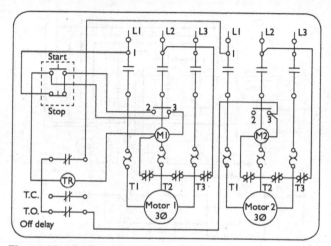

Figure 19-12 Electrical Diagram Two Motors with Off Timer for Second Motor

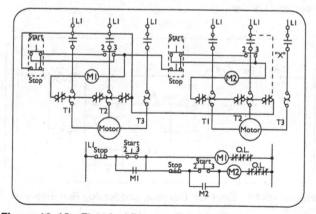

Figure 19-13 Electrical Diagram Two Motors for Conveyor. Motor M-2 Can Start Only after Motor M-1 Starts.

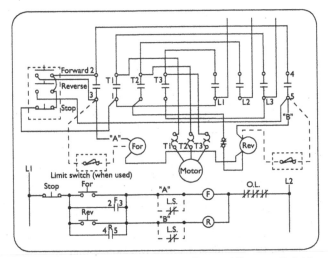

Figure 19-14 Electrical Diagrams, Forward Reverse Point to Point and Ladder

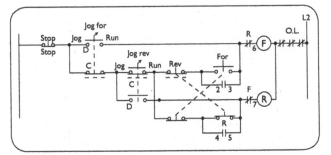

Figure 19-15 Electrical Diagram, Ladder Jog Run Selector Switch

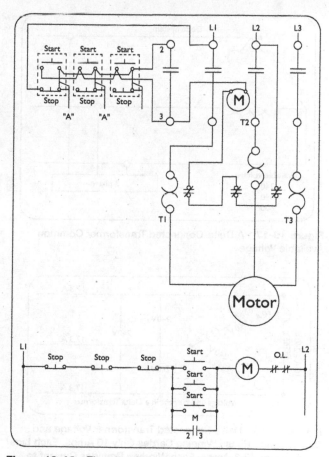

Figure 19-16 Electrical Diagram for Multiple Start-Stop Stations Point to Point and Ladder

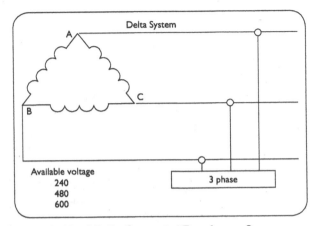

Figure 19-17 A Delta-Connected Transformer Common
Available Voltage

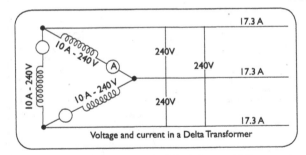

Figure 19-18 Delta-Connected Transformer Voltage and
Current. Each Phase Winding Carries Only 10 Amps. Each Line
Lead Carries 17.3 Amps. Each Winding Provides Current to
Each Lead.

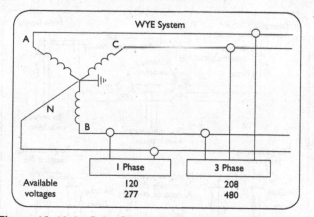

Figure 19-19 A   Delta-Connected Transformer Available Voltages

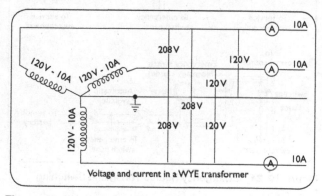

Figure 19-20   Wye-Connected Transformer, Voltage and Current

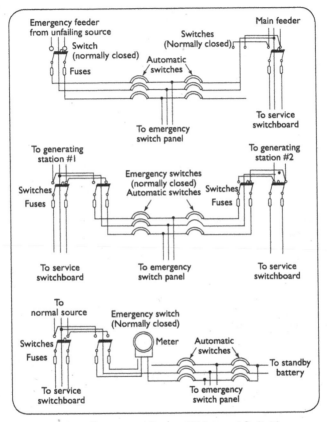

Figure 19-21 Emergency Source Wiring and Switching Methods

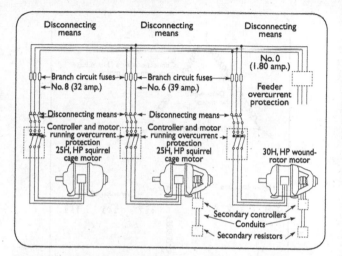

Figure 19-22 Electrical Diagram for Three-Motor Installation Showing Feeder and Branch Circuit Wiring

Top diagram: Two sources of power to emergency panel, one source to switchboard. Middle diagram: One source to each of two switchboards, two sources to emergency panel. Lower: One source to switchboard, two sources to emergency panel. One source is connected to stand-by battery supply.

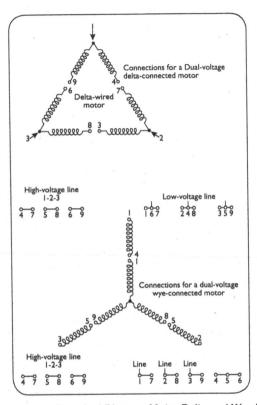

Figure 19-23 Electrical Diagram Motor Delta and Wye Wiring

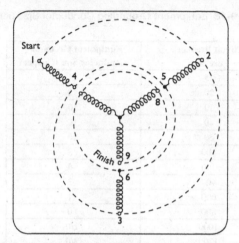

Figure 19-24 Electrical Diagram Motor Wiring Spiral

Table 19-14 Conductor Ampacity Aluminum, RH, RHH, RHW, THHW, THW, THWN, THHN, XHHW

Copper Wire Size in AWG	Service Rating in Amperes
4	100
3	110
2	125
1	150
1/0	175
2/0	200

Table 19-15 Equipment Grounding Conductor Size per OCPD Size

Circuit Breaker/ Fuse Size	Equipment Grounding Conductor Size (Copper)
15	14
20	12
30	10
40	10
60	10
100	8
200	6
400	3
600	1
800	0
1000	2/0
1200	3/0
1600	4/0
2000	250 KCMIL
2500	350 KCMIL
3000	400 KCMIL
4000	500 KCMIL
5000	700 KCMIL
6000	800 KCMIL

Table 19-16 Maximum Number of THHN, THWN, THWN-2 Insulated Conductors in Various Rigid Metal Conduit Sizes (As Specified in Chapter 9 of The NEC)

Insulation THHH THWN THWN-2 AWG	½ RMC	¾ RMC	1 RMC	1¼ RMC	1½ RMC	2 RMC	2½ RMC	3 RMC	3½ RMC	4 RMC
14	13	22	36	63	85	140	200	309	412	531
12	9	16	26	46	62	102	146	225	301	387
10	6	10	17	29	39	64	92	142	189	244
8	3	6	9	16	22	37	53	82	109	140
6	2	4	7	12	16	27	38	59	79	101
4	1	2	4	7	10	16	23	36	48	62
3	1	1	3	6	8	14	20	31	41	53
2	1	1	3	5	7	11	17	26	34	44
1	1	1	1	4	5	8	12	19	25	33
1/0		1	1	3	4	7	10	16	21	27
2/0		1	1	2	3	6	8	13	18	23
3/0		1	1	1	2	5	7	11	15	19
4/0		1	1	1	1	4	6	9	12	16
250 KCMIL			1	1	1	3	5	7	10	13
300 KCMIL			1	1	1	3	4	6	8	11
350 KCMIL			1	1	1	2	3	5	7	10
400 KCMIL				1	1	2	3	5	7	8
500 KCMIL				1	1	1	2	4	5	7
600 KCMIL				1	1	1	1	3	4	6
700 KCMIL				1	1	1	1	3	4	5
750 KCMIL				1	1	1	1	3	4	5

Table 19-17 Maximum Number of Conductors Insulation Type TW, THHW, THW, THW-2 in Various Sizes of Conduit

Insulation TW, THHW, THW, THW-2	Conduit Size	½	¾	1	1¼	1½	2	2½	3
	14	9	15	25	44	59	98	140	216
	12	7	12	19	33	45	75	1-7	165
	10	5	9	14	25	3456	80	123	164
	8	3	5	8	14	19	31	44	68
	6	1	3	5	8	11	18	27	41
	4	1	1	3	6	8	14	20	31
	3	1	1	3	5	7	12	17	26
	2	1	1	2	4	6	10	14	22
	1	1	1	1	3	4	7	10	15
	1/0	0	1	1	2	3	6	8	13
	2/0	0	1	1	2	3	5	7	11
	3/0	0	1	1	1	2	4	6	9
	4/0	0	0	1	1	1	3	5	8
	250	0	0	1	1	1	3	4	6
	300	0	0	1	1	1	2	3	5
	350	0	0	0	1	1	1	3	5
	400	0	0	0	1	1	1	3	4
	500	0	0	0	1	1	1	2	3
	600	0	0	0	1	1	1	1	3
	700	0	0	0	0	1	1	1	2
	750	0	0	0	0	1	1	1	2

Table 19-18 Metal Box Maximum Conductor Fill

Box Size/ Type	In³ Cap.	18 AWG	16 AWG	14 AWG	12 AWG	10 AWG	8 AWG	6 AWG
4 × 1 ¼ R/0	12	8	7	6	5	5	5	2
4 × 1½ R/0	15.5	10	8	7	6	6	5	3
4 × 2½ R/0	21.5	14	12	10	9	8	7	4
4 × 1 ¼ Sq.	18.0	12	10	9	8	7	6	3
4 × 1/12 Sq.	21.0	14	12	10	9	8	7	4
4 × 2 1/8 Sq.	30.0	20	17	15	13	12	10	6
4 11/16 × 1 ¼ Sq.	25.5	17	14	12	11	10	8	5
4 11/16 × 1 ½ Sq.	29.5	19	16	14	13	11	9	5
4 11/16 × 2 1/8 Sq.	42.0	28	24	21	18	16	14	8
3 × 2 × 1 ½ Device	7.5	5	4	3	3	3	2	1
3 × 2 × 2 Device	10.0	6	5	5	4	4	3	2
3 × 2 × 2 ¼ 1Device	10.5	7	6	5	4	4	3	2
3 × 2 × 2 ½ Device	12.5	8	7	6	5	5	4	2
3 × 2 × 2 ¾ Device	14.0	9	8	7	6	5	4	2
3 × 2 × 3 ½ Device	18.0	12	10	9	8	7	6	3
4 × 2 1/8 × 1 ½ Device	10.3	6	5	5	4	4	3	2

(continued)

Table 19-18 (Continued)

Box Size/ Type	In³ Cap.	18 AWG	16 AWG	14 AWG	12 AWG	10 AWG	8 AWG	6 AWG
4 × 2 1/8 × 1 7/8 Device	13.0	8	7	6	5	5	4	2
4 × 2 1/8 × 2 1/8 Device	14.5	9	8	7	6	5	4	2
3 ¾ × 2 × 2 ½ Masonary/gang	14.0	9	8	7	6	5	4	2
3 ¾ × 2 × 3 ½ Masonary/gang	21.0	14	12	10	9	8	7	4
FS- minimum depth 1 ¾ single cover/gang	13.5	9	7	6	6	5	4	2
FD- minimum depth 2 3/8 single cover/gang	18.0	12	10	9	8	7	6	3
FD- minimum depth 1 ¾ multiple cover/gang	18.0	12	10	9	10	7	6	3
FD- minimum depth 2 3/8 multiple cover/gang	24.0	16	13	12	9	8	8	4

Table 19-19 Box Fill for Various Conductor Sizes

Conductor AWG	Free Space for Each Wire in In³
18	1.5
16	1.75
14	2.0
12	2.25
10	2.5
8	3.0
6	5.0

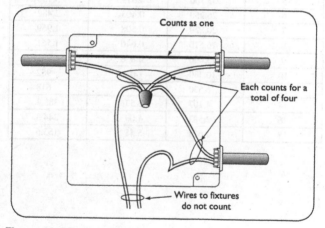

Figure 19-25 Box Fill Conductor Counts

Table 19-20 Electrical Conductor Copper Data

Conductor Size AWG	Area in CMIL	Ohms per 1000 ft	Feet per ohm
4/0	211,600	0.0500	20,010
3/0	167,800	0.0630	15,870
2/0	133,100	0.0792	12580
1/0	105,500	0.01002	9,980
1	83,690	0.1264	7,914
2	66,370	0.1593	6,276
3	52,640	0.2009	4,977
4	41,740	0.2533	3,947
5	33,100	0.3195	3,130
6	26,250	0.4030	2,482
7	20,820	0.508	1,969
8	16,510	0.640	1,561
9	13,090	0.808	1,238
10	10,380	1.018	982
12	6,530	1.616	618
14	4,107	2.575	388.3
16	2,583	4.09	244.3
18	1,624	6.51	153.6

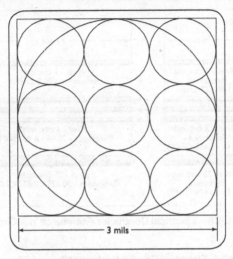

Figure 19-26 Comparison of 9 Circular to 9 Square Mils

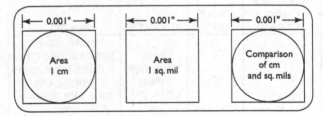

Figure 19-27 Electrical Conductor Comparison between 1 Circular Mil, and 1 Square Mil

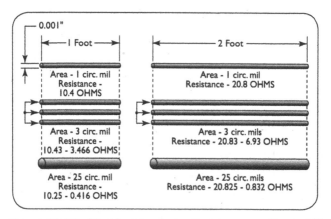

Figure 19-28 Electrical Conductor Resistance per Circular Mil

Table 19-21 Electrical Conductor Ampacity

Not More Than Three Wires in Conduit 30 °C/86 °F	60 °C/ 140 °F	75 °C/167 °F	90 °C/194 °F
Copper Size AWG/ KCMIL	TW, UF	RHW, THHW, THW, THWN, XHHW, USE, ZW	TBS, SA, SIS, FEP, FEPB, MI, RHH, RHW-2, THHN, THHW, THW-2, THWN-2, USE-2, XHH, XHH, XHHW-2, ZW-2
18	-0-	-0-	14
16	-0-	-0-	18
14	15	20	25
12	20	25	30
10	30	35	40
8	40	50	55

Table 19-21 (Continued)

Not More Than Three Wires in Conduit 30 °C/86 °F	60 °C/ 140 °F	75 °C/167 °F	90 °C/194 °F
6	55	65	75
4	70	85	95
3	85	100	115
2	95	115	130
1	110	130	145
1/0	125	150	170
2/0	145	175	195
3/0	165	200	225
4/0	195	230	260
250 KCMIL	215	255	290
300 KCMIL	240	285	320
350 KCMIL	260	310	350
400 KCMIL	280	335	380
500 KCMIL	320	380	430
600 KCMIL	350	420	475
700 KCMIL	385	460	520
750 KCMIL	400	475	535
800 KCMIL	410	490	555
900 KCMIL	435	520	585
1000 KCMIL	455	545	615

Table 19-22 Resistance for Various Metals in Ohms per Circular Mil Foot

Metal	Ohms per Circular-Mil Foot
Aluminum	17
Copper, annealed	10.4
Iron	60.2
Nichrome wire	660.0
Platinum	59.5
Silver	9.9
Tungsten	33.1
Zinc	36.7
Brass	42.1

Table 19-23 Conductivity of Various Metals as a Percentage of Silver

Metal	Conductivity as a % of Sliver
Silver	100
Copper	98
Gold	78
Aluminum	61
Zinc	30
Platinum	17
Iron	16
Tin	9

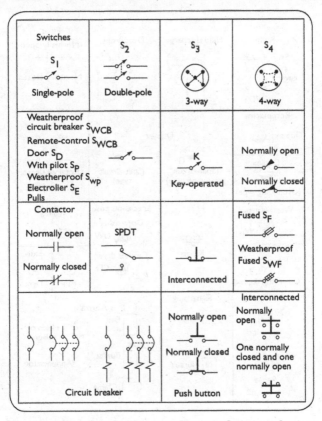

Figure 19-29 Electrical Diagram Symbols Switches, Contactors, and Circuit Breakers

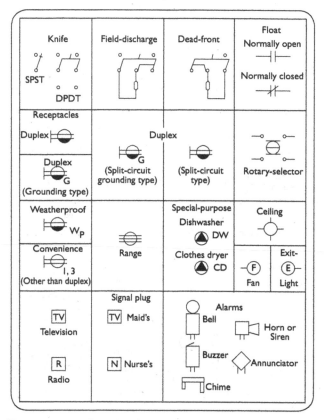

Figure 19-30 Electrical Symbols, Receptacles, and Alarms

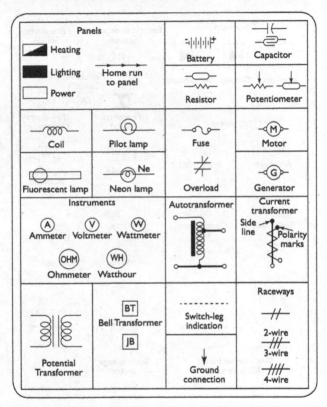

Figure 19-31 Electrical Symbols, Coils, Meters Raceways

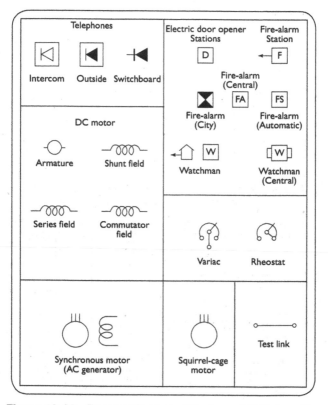

Figure 19-32 Electrical Symbols, Telephone, DC Motors, Fire Alarm

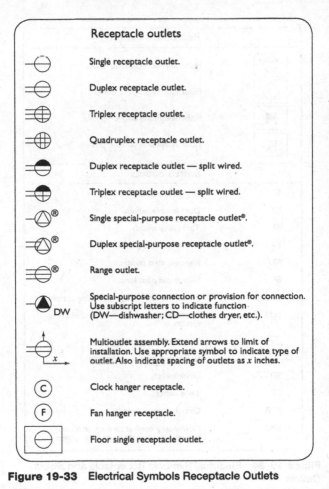

Figure 19-33 Electrical Symbols Receptacle Outlets

	Receptacle outlets
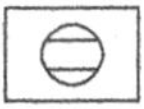 ⊖	Floor duplex receptacle outlet.
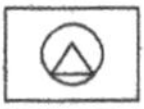 ⊘	Floor special-purpose outlet.
 ◄	Floor telephone outlet—public.
 ◁	Floor telephone outlet—private.

	Switch outlets
S	Single-pole switch.
S2	Double-pole switch.
S3	Three-way switch.
S4	Four-way switch.
SK	Key-operated switch.
SP	Switch and pilot lamp.
SL	Switch for low-voltage switching system.
SLM	Master switch for low-voltage switching system.
 ⊖S	Switch and single receptacle.
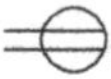 ⊜S	Switch and double receptacle.
SD	Door switch.
ST	Time switch.
SCB	Circuit-breaker switch.
SMC	Momentary contract switch or push-button for other than signaling system.

Figure 19-34 Electrical Symbols Receptacle and Switch Outlets

Circuiting

	Wiring concealed in ceiling or wall.
———————	Wiring concealed in floor.
– – – – –	Wiring exposed. Note: Use heavyweight line to identify service and feeders. Indicate empty conduit by notation CO (conduit only).
....	

3 wires

————————— 2 ——→ 1

——//———

Branch-circuit home run to panelboard. Number of arrows indicates number of circuits. (A numeral at each arrow may be used to identify circuit number.) Note: Any circuit without further identification indicates two-wire circuit. For a greater number of wires, indicate with cross lines.

4 wires, etc.

——////———

Unless indicated otherwise, the wire size of the circuit is the minimum size required by the specification.

Identify different functions of wiring system—for example, signaling system by notation or other means.

————————o · Wiring turned up.

————————● Wiring turned down.

Lighting Outlets

Ceiling   Wall

◯	—◯	Surface or pendant incandescent, mercury vapor, or similar lamp fixture.
Ⓡ	—Ⓡ	Recessed incandescent, mercury vapor, or similar lamp fixture.

Figure 19-35   Electrical Symbols Circuiting

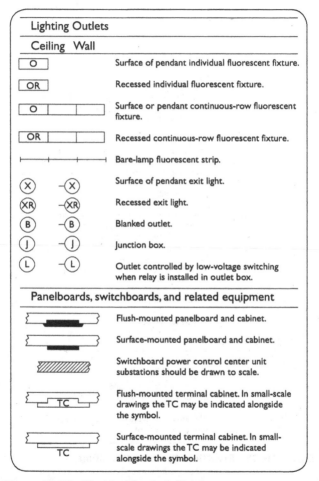

Figure 19-36 Electrical Symbols Lighting

Table 19-24 Sound Level in dB for Residences, Retail Store, Office, Factory Floor

Area	Sound Level in dB
Residential	30 to 45
Retail store	45 to 55
Office	50 to 75
Factory floor	75 to 95

Table 19-25 Electrical Transformer Sound Power Level by Size

Transformer kVA rating	Typical Sound Level in dB
10 and smaller	40
11 to 50	45
51 to150	50
151 to 300	55
301 to 500	60

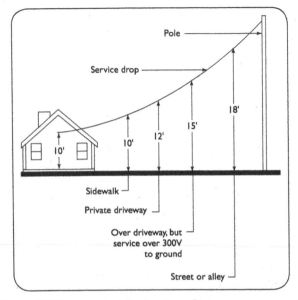

Figure 19-37 Electrical Service Drop Clearance

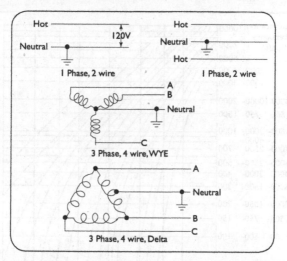

Figure 19-38 Electrical Transformer Wye and Delta Grounding

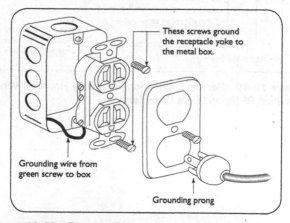

Figure 19-39 Receptacle and Metal Box Bonding

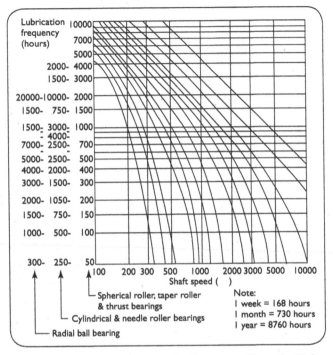

Figure 19-40 Bearing Lubrication Frequency Hours for Various Shaft RPMs and Bore Diameters

Table 19-26 Electrical Motor Lubrication for Motors with Ball Type Bearings By Duty and Horsepower

Type of Service	Example	HP Range ½ to 7.5	HP Range 10 to 40	HP Range 50 to 200
Light	1 hr run time per day	10 years	7 years	5 years
Normal	Machine tools, fans, pumps	7 years	5 years	3 years
Heavy	Continuous critical machines	4 years	2 years	1 year
Severe	Dirty, vibrating, high ambient Multiple starts, plugging operations	9 months	4 months	4 months

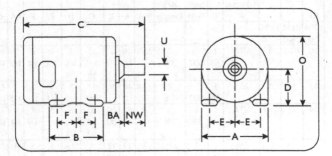

Figure 19-41 NEMA Motor Frame Dimension Designations

Table 19-27 NEMA Electrical Motor T-Frame Dimensions 1.5 to 7.5 Hp

HP			T-Frame #	U-Shaft OD	Key way	Key Length	N-W	A max	B max	C	D	E	F	BA	O
RPM 3600	RPM 1800	RPM 1200			W/d										
1 1/2	1	3/4	143	7/8	3/16-3/32	1 3/8	2 1/4	7	6	12 5/8	3 3/4	2 3/4	2	2 1/4	7
1-1															
1 ½-2															
5	3	11 1/2													
7 1/2	5	2	184	1 1/8	1/4-1/8	1 3/4	2 3/4	9	7 1/2	13 1/2	4 3/4	3 1/2	2 3/4	2 3/4	9

Table 19-28 NEMA Electrical Motor T-Frame Dimensions

HP			T-Frame #	U	Key Seat W/D	Key Length	NW	A	B	C	D	E	F	BA	0
RPM 3600	RPM 1800	RPM 1200													
10	7 1/2	3	213	1 3/8	5/16-3/32	2 3/8	3 3/8	10 1/2	7 1/2	15 13/16	5 1/4	4 1/4	2 3/4	3 1/2	10 1/2
15	10	5	215	1 3/8	5/16-3/32	2 3/8	3 3/8	10 1/2	9	17 1/16	5 1/4	4 1/4	3 1/2	3 1/2	10 1/2
20	15	7 ½	254	1 5/8	1 5/8	2 7/8	4	12 1/2	10 1/2	20 1/2	6 1/4	5	4 1/4	4 1/4	12 1/2
25	20	10	256	1 5/8	1 5/8	2 7/8	4	12 1/2	12 1/2	22 1/4	6 1/4	5	5	4 1/4	12 1/2
	25	15	284T	1 7/8	1 7/8	3 1/4	4 3/8	12 1/2	12 1/2	23 5/16	7	5 1/2	5 1/2	4 3/4	14
30	25	15	284TS	1 5/8	1 5/8	1 7/8	3 1/4	12 1/2	12 1/5	22	7	5 1/2	5 1/2	4 3/4	14

Table 19-28 *(Continued)*

HP			T-Frame #	U	Key Seat W/D	Key Length	NW	A	B	C	D	E	F	BA	0
	30	20	286T	1 7/8	1 7/8	3 1/4	4 3/8	14	14	24 7/8	7	5 1/2	5 1/2	4 3/4	14
40	30	20	286TS	1 5/8	1 7/8	1 7/8	1 7/8	14	14	23 1/2	7	5 1/2	5 1/2	4 3/4	14
	40	25	324	2 1/8	2 1/8	3 7/8	5 1/4	14	14	26 1/2	8	6 1/4	5 1/4	5 1/4	16
50	40	25	324TS	1 7/8	1 7/8	2	3 3/4	14	14	24 5/8	8	6 1/4	5 1/4	5 1/4	16
	50	30	326	2 1/8	2 1/8	3 7/8	5 1/4	15 1/2	15 1/2	27 3/4	8	6 1/4	6	5 1/4	16
60	50	30	326TS	1 7/8	1 7/8	1/2	3 3/4	16	15 1/2	26 1/8	8	6 1/4	6	5 1/4	16
	60	40	364	2 3/8	2 3/8	4 1/4	5 7/8	18	15 1/4	28 3/4	9	7	5 5/8	5 7/8	
75	60		364TS	1 7/8	1 7/8	2	3 3/4	18	15 1/4	26 9/16	9	7	5 5/8	5 7/8	
	75	80	365	2 3/8	2 3/8	4 1/4	5 7/8	18	16 1/4	29 3/4	9	7	6 1/8	5 7/8	
100	75		365TS	1 7/8	1 7/8	2	3 3/4	18	16 1/4	27 9/16	9	7	6 1/8	5 7/8	

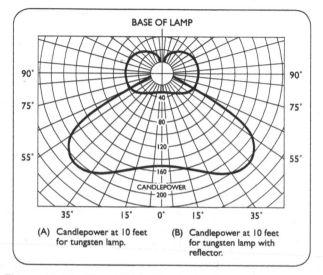

BASE OF LAMP

90° 90°

75° 75°

40

80

120

55° 55°

160

CANDLEPOWER

200

35°   15°   0°   15°   35°

(A) Candlepower at 10 feet
   for tungsten lamp.

(B) Candlepower at 10 feet
   for tungsten lamp with
   reflector.

Figure 19-42 Electric Light Distribution Curve

As can be seen in the curve shown in Figure 19-42, light output distribution is not circular, but follows a uniform pattern that is symmetrical for both sides of the pattern. The pattern for each lamp fixture is different.

Table 19-29 Percent Light Reflected from Wall and Ceilings for Various Colors with a Uniform Surface

Surface	Class	Color	Percent Light Reflected
Paint		White	81
Paint	Light	Ivory	79
Paint		Cream	74
Paint		Buff	63
Paint	Medium	Light green	63
Paint		Light green	58
Paint		Tan	48
Paint		Dark green	26
Paint		Olive green	17
Wood	Dark	Light oak	32
Wood		Dark oak	13
Wood		Mahogany	8
Cement		Natural	25
Brick		Red	13

Table 19-30 Wire Size Required to Maintain a 2-V Drop for Various Wattages and Distances for 120-V Circuits

Load	Amps at 120 V	30 FT	40 FT	50 FT	60 FT	70 FT	80 FT	90 FT	100 FT	110 FT	120 FT	130 FT	140 FT	150 FT	160 FT	170 FT	180 FT	190 FT	200 FT
500	4.2	14	14	14	14	14	14	12	12	12	12	12	12	12	10	10	10	10	10
600	5.0	14	14	14	14	14	12	12	12	12	10	10	10	10	10	10	10	8	8
700	5.8	14	14	14	14	12	12	12	10	10	10	10	10	10	18	8	8	8	8
800	6.7	14	14	14	12	12	10	10	10	10	10	8	8	8	8	8	8	8	8
900	7.5	14	14	14	12	12	10	10	10	10	8	8	8	8	8	8	6	8	6
1000	8.3	14	14	12	12	12	10	10	10	10	8	8	8	8	8	8	6	6	6
1200	10.0	14	12	12	12	10	8	8	8	8	8	8	8	6	6	6	6	6	6
1400	11.7	14	12	12	10	10	8	8	8	8	8	6	6	6	6	6	6	6	6
1600	13.3	12	12	12	10	8	8	8	8	6	6	6	6	6	6	4	4	4	4
1800	15.0	12	12	10	10	8	8	8	6	6	6	6	6	6	4	4	4	4	4
2000	16.7	12	10	10	8	8	8	6	6	6	6	6	6	4	4	4	4	4	4

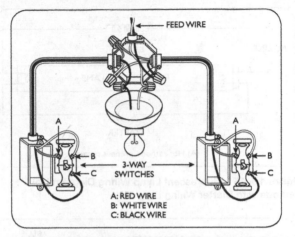

Figure 19-43 Wiring Diagram for Two Three-Way Switches Controlling a Single Light Wiring Diagram

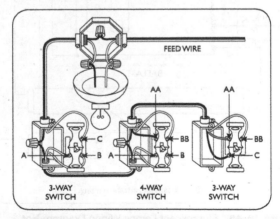

Figure 19-44 One Lamp Controlled by Any One of Three Switches Wiring Diagram

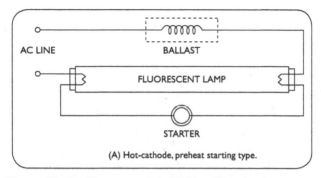

Figure 19-45 Fluorescent Lamp Wiring Diagram Hot
Cathode with Starter Wiring Diagram

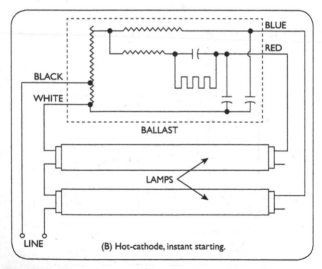

Figure 19-46 Fluorescent Lamps Wiring Diagram Hot
Cathode Instant-Start Wiring Diagram

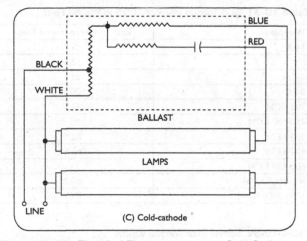

Figure 19-47 Electrical Fluorescent Lamps Cold Cathode Wiring Diagram

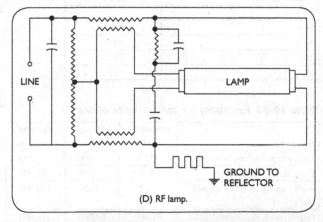

Figure 19-48 Electrical Fluorescent Rf Lamp Wiring Diagram

Table 19-31 Bolt Torque per Wire Size in Pound Feet

Wire Size	Bolt Torque in Pound Feet
14-8	6.25
6-4	12.5
3-1	21
0-2/0	29
200 KCM	37.5
250-300 KCM	50
400 KCM	62.5
500 KCM	62.5
600-750 KCM	75
800-1000 KCM	83.25
1250-2000 KCM	83.3

Table 19-32 Screw Size Torque Values

Screw Size Across the Flats in Inches	Torque in Pound Feet
1/8 inch	4.2
5/34	8.3
3/16	15
7/32	23.25
¼	42

Table 19-33 Resistivity for Various Types of Soil

Soil Type	Resistivity Average	Ohms Minimum	Per CM Maximum
Fills, ashes, cinders, brine waste	2370	590	7000
Clay, shale, gumbo, loam	4060	340	16,300
With varying amounts of sand and gravel	15,800	1020	136,000
Gravel, sand, stones, with little clay or loam	94,000	59,000	458,000

Table 19-34 Soil Resistivity Changes by Moisture Content

Moisture Content Percent by Weight	Resistivity in Ohms per CM Top Soil	Sandy Loam
2.5%	1000×10^6	1000×10^6
0	250,000	150,000
5%	165,000	43,000
10	53,000	18,500
15	19,000	10,500
20	12,000	6,300
25	6,400	4,200

Table 19-35 Soil Resistivity Changes by Salt Content Change

Added Salt Percent by Weight of Moisture	Resistivity in Ohms CM
0	10,700
0.1	1,800
1.0	460
5	190
10	130
20	100
For sandy loam soil, with 15% moisture at 63 °F	

Table 19-36 Mercury Vapor Lamp Open Circuit Voltage

Wattage	Voltage Range
50	215–270
75	220–225
100	225–285
125	230–290
175	200–290
250	210–285
400	210–285
1000	385–465

Table 19-37 Conductor Copper Vertical Support Spacing

Conductor Size AWG	Support Maximum Spacing
18- 8	100
6-1/0	100
2/0-4/0	80
4/0- 350 KCMIL	60
350-500 KCMIL	50
500-750 KCMIL	40
750 + KCM	35

Table 19-38 Conduit Inside Diameter in Inches for PVC, RMC, and EMT

Conduct Trade Size	Conduit Material PVC	Conduit Material RMC	Conduit Material EMT
¾	0.804	0.836	0.824
1	1.029	1.063	1.049
1 ¼	1.360	1.394	1.380
1 ½	1.590	1.624	1.610
2	2.047	2.083	2.067
2 ½	2.445	2.498	2.731
3	3.042	3.090	3.356
3 ½	3.521	3.570	3.834
4	3.998	4.050	4.334
5	5.016	6.093	
6	6.031		

Table 19-39 Conductor Mega-ohm Value Temperature Correction Factor

Ambient Temperature	Insulation Type XHHW/RHH/RHW/ USE	Insulation Type THHN
50 °F	0.73	0.56
55 °F	0.86	0.75
60 °F	1.00	1.0
65 °F	1.17	1.34
70 °F	1.36	1.79
75 °F	1.59	2.40
80 °F	1.86	3.21
85 °F	-0-	4.30

Conductor Insulation Resistance (IR) Mega-Ohm Reading

Formula IR in mega-ohms per 1000 ft = $(L \times R \times F) \times 1000$
Where IR = insulation resistance in mega-ohms per 1000 ft.

L = conductor length, R = mega-ohm reading, F = temperature correction factor listed above for the type of insulation materials

Table 19-40 Insulated Copper Conductor Weight per 1000 Ft

AWG Size	Weight per 1000 ft
14	18
12	26
10	40
8	66
6	99
4	152

Table 19-40 *(Continued)*

AWG Size	Weight per 1000 ft
2	234
1	299
1/0	371
2/0	461
3/0	574
4/0	717
250 KCMIL	850
300 KCMIL	1011
350 KCMIL	1173
400 KCMIL	1333
500 KCMIL	1653
600 KCMIL	1985
750 KCMIL	2462
1000 KCMIL	3254

Table 19-41 Wire Lubricant Quantity Recommended for 100 ft of Conduit

Conduit Trade Size in Inches	Quantity of Wire Lubricant in Gallons per 100 ft of Conduit
2	0.3 gal
3	0.45 gal
4	0.6 gal
6	0.9 gal

Formula quantity = $0.0015 \times L \times D$. Where L = length of conduit, D = inside diameter of the conduit in inches

Table 19-42 Conductor Jamming Ratio for Three or More Conductors

Ratio	Risk
2.4 or less	Very small
2.4 to 2.5	Small
2.5 to 2.6	Moderate
2.6 to 2.9	Significant
2.9 to 3.0	Moderate
3.0 to 3.2	Small
3.2 and above	Very small

Jamming ratio is conduit inside diameter/conductor outside diameter.

Table 19-43 Conductor Dynamic Friction by Conductor Insulation and Conduit Type

Insulation Type	EMT Conduit	PVC Conduit
THHW/THNW	.28	.24
XHHW/USE	.28	.24
RHH/RHW/XLPE	.25	.14

Table 19-44 Copper Conductor Maximum Pulling Tension by AWG. Should Not Exceed 0.008 Pounds per Circular Mill, or Manufacturer's Specifications, Whichever Is Smaller

AWG	Maximum Pulling Tension in Pounds
14	32.9
12	83
10	132
8	167
6	210
4	334
2	531

(continued)

AWG	Maximum Pulling Tension in Pounds
1	670
1/0	845
2/0	1065
3/0	1342
4/0	1693
250 KCM	2000
300 KCM	2400
350 KCM	2800
400 KCM	3200
450 KCM	3600
500 KCM	4000
550 KCM	4400
600 KCM	4800
650 KCM	5200
700 KCM	5600
750 KCM	6000
800 KCM	6400
900 KCM	7200
1000 KCM	8000
1200 KCM	9600

Table 19-45 Conductor Minimum Bend Radius, Insulation Type RHH/RHW/RHHW-2

Conductor Size	Minimum Bend Radius
350 KCMIL and smaller	4 times conductor OD
400 – 1750 KCMIL	5 times conductor OD
2000 KCMIL and up	6 times conductor OD

Table 19-46 Conductor Minimum Bend Radius, Insulation Type THHW/THW-2/RHH/RHW-2

Conductor Size	Minimum Bend Radius
500 KCMIL and smaller	4 times conductor OD
600 – 1000 KCMIL	5 times conductor OD

Table 19-47 Conductor Minimum Bend Radius, Insulation Type XHHW/XHHW-2/XHH

Conductor Size	Minimum Bend Radius
500 KCMIL	4 times conductor OD
600 – 2000 KCMIL	5 times conductor OD

Table 19-48 Conductor # 12 to 4/0 AWG and Circular Mil Values

Conductor Size in AWG	Conductor Size in Circular Mils
12	6530
10	10,380
8	16,510
6	26,240
4	41,740
2	52,620
1	66,360
1/0	83,690
2/0	105,600
3/0	133,100
4/0	167,800

Table 19-49 Conductor # 250 KCMIL to 1,000 KCMIL and Circular Mil Values

Conductor Size in KCMIL	Conductor Size in Circular Mils
250	250,000
300	300,000
350	350,000
400	400,000
500	500,000
600	600,000
750	750,000
800	800,000
1,000	1,000,000

Table 19-50 Twisted Pair Cable Categories

Category	Maximum Rate of Data Transfer	Application
1	1 Mbps	Voice
2	4 Mbps	Token ring
3	16 Mbps	10 Base T Ethernet
4	20 Mbps	16 Mbps Token ring
5	100 Mbps (2 pairs/four wires)	100 base T Ethernet
5e	1,000 Mbps (4 pairs/ eight wires)	Gigabit Ethernet
6	1,000 Mbps (4 pairs/eight wires)	Gigabit Ethernet
6a	10,000 Mbps	10 Gigabit Ethernet
7	10,000 Mbps	10 Gigabit Ethernet

Table 19-51 Pin Connections for Twisted Pair Cable

Pin Number	Function	EIA/TIA 568A	EIA/TIA 568B and AT & T 258 A
#1	+ transmit	White/green	White/orange
#2	− Transmit	Green	Orange
#3	+ Receive	White/orange	White/green
#4	Unused	Blue	Blue
#5	Unused	White/blue	White/blue
#6	− Receive	Orange	Green
#7	Unused	White/brown	White/brown
#8	Unused	Brown	Brown

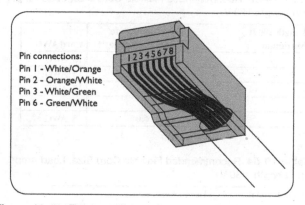

Pin connections:
Pin 1 - White/Orange
Pin 2 - Orange/White
Pin 3 - White/Green
Pin 6 - Green/White

Figure 19-49 Ethernet RJ-45 Connector Pin Sequence

One can determine what is what by holding the retaining clip on top. Pins will be on the bottom, pin with # one on the left, with color code as shown.

Table 19-52 Crossover Cable Pin Sequence and Conductor Color Code

Pin #	Connector A	Connector B
1	White/green	White/orange
2	Green	Orange
3	White/orange	White/green
4	Blue	Blue
5	White/blue	White/blue
6	Orange	Green
7	White/brown	White/brown
8	Brown	Brown

Table 19-53 Recommended Flexible Cord Size, Load Amps, and Length 120 V

Length 100 ft Maximum	Load Nameplate Amperage Range	Cord AWG
	0 to 3.4A	16 AWG
	3.5 to 5.0 A	14 AWG
	5.1 to 7.0 A	12 AWG
	7.1 to 12.0	10 AWG
	12.1 to 16.0	8 AWG

Table 19-54 Recommended Flexible Cord Size, Load Amps, and Length, 230 V

Length 100 ft Maximum	Load Nameplate Amperage Range	Cord AWG
	0 to 3.4	18 AWG
	3.5 to 5.0	16 AWG
	5.1 to 7.0	14 AWG
	7.1 to 12.0	12 AWG
	12.1 to 16.0	10 AWG

Table 19-55 Flexible Cords Type C

Flexible Cord and Cable	Feature Included
Type letter	C
Voltage rating	330 – 600
AWG range	18 – 16 and 14 – 10
Number of conductors	2 or more
Location Dry	Y
Location wet	N
Location hazardous	N
Usage not hard	Y
Usage Hard	N
Usage extra hard	N
Usage attached to appliance	N
Usage pendant or portable	Y
Usage refrigerators room AC as permitted	N
Usage electric vehicle charging	N
Usage elevator	Y
Insulation type	Thermoset or thermoplastic
Insulation thickness range	3- to 45 mils
Outer covering	N
Outer covering flame retardant	N
Outer covering moisture resistant	N
Outer braid on each conductor	Cotton

Table 19-56 Flexible Cords Type E

Flexible Cord and Cable	Feature Included
Type letter	E
Voltage rating	300 – 600
AWG range	20 – 2
Number of conductors	2 or more
Location dry	Unclassified

(continued)

Table 19-56 (Continued)

Flexible Cord and Cable	Feature Included
Location wet	Unclassified
Location hazardous	Unclassified
Usage not hard	Unclassified
Usage hard	N
Usage extra hard	N
Usage attached to appliance	N
Usage pendant or portable	N
Usage refrigerators room AC as permitted	N
Usage electric vehicle charging	N
Usage elevator	Y
Insulation type	Thermoset
Insulation thickness range	20-30-45-60 mils
Outer covering	Y
Outer covering flame retardant	Y
Outer covering moisture resistant	Y
Outer braid on each conductor	Flexible nylon jacket

Table 19-57 Flexible Cords Type EO

Flexible Cord and Cable	Feature Included
Type letter	EO
Voltage rating	300 – 600
AWG range	20 – 2
Number of conductors	2 or more
Location dry	Unclassified
Location wet	Unclassified
Location hazardous	Y
Usage not hard	Unclassified
Usage hard	Unclassified
Usage extra hard	Unclassified
Usage attached to appliance	N

Table 19-57 (Continued)

Flexible Cord and Cable	Feature Included
Usage pendant or portable	N
Usage refrigerators room AC as permitted	N
Usage electric vehicle charging	N
Usage elevator	Y
Insulation type	Thermoset
Insulation thickness range	20-30-45-60 mils
Outer covering	Y
Outer covering flame retardant	Y
Outer covering moisture resistant	Y
Outer braid on each conductor	Cotton

Table 19-58 Flexible Cords Type ETP

Flexible Cord and Cable	Feature Included
Type letter	ETP
Voltage rating	300-600
AWG range	20-14 300 V & 10-2 600 v & 12 300V
Number of conductors	Varies
Location dry	N
Location wet	N
Location hazardous	Y
Usage not hard	N
Usage hard	N
Usage extra hard	N
Usage attached to appliance	N
Usage pendant or portable	N
Usage refrigerators room AC as permitted	N
Usage electric vehicle charging	N
Usage elevator	Y
Insulation type	Rubber or Thermoplastic

(continued)

Table 19-58 *(Continued)*

Flexible Cord and Cable	Feature Included
Insulation thickness range	30 – 45 mils
Outer covering	Y
Outer covering flame retardant	N
Outer covering moisture resistant	N
Outer braid on each conductor	N

Table 19-59 Flexible Cords Type ETT

Flexible Cord and Cable	Feature Included
Type letter	ETT
Voltage rating	300 – 600
AWG range	20 –14 300 V & 10 – 2 600 v & 12 300V
Number of conductors	Varies
Location dry	N
Location wet	N
Location hazardous	Y
Usage not hard	N
Usage hard	N
Usage extra hard	N
Usage attached to appliance	N
Usage pendant or portable	N
Usage refrigerators room AC as permitted	N
Usage electric vehicle charging	N
Usage elevator	Y
Insulation type	Rubber or Thermoplastic
Insulation thickness range	30 – 45 mils
Outer covering	Thermoplastic
Outer covering flame retardant	N
Outer covering moisture resistant	N
Outer braid on each conductor	N

Table 19-60 Flexible Cords Type EV

Flexible Cord and Cable	Feature Included
Type letter	EV
Voltage rating	600
AWG range	18-500
Number of conductors	2 or more + grounding
Location dry	N
Location wet	Y
Location hazardous	N
Usage not hard	N
Usage hard	N
Usage extra hard	Y
Usage attached to appliance	N
Usage pendant or portable	N
Usage refrigerators room AC as permitted	N
Usage electric vehicle charging	Y
Usage elevator	N
Insulation type	Thermoset with optional nylon
Insulation thickness range	30-45-60-80-95/Nylon 20-30-45-60-75 mils
Outer covering	Thermoset
Outer covering flame retardant	N
Outer covering moisture resistant	N
Outer braid on each conductor	Optional

Table 19-61 Flexible Cords Type EVJ

Flexible Cord and Cable	Feature Included
Type letter	EVJ
Voltage rating	300
AWG range	18-12
Number of conductors	2 or more + grounding

(continued)

Table 19-61 *(Continued)*

Flexible Cord and Cable	Feature Included
Location dry	N
Location wet	Y
Location hazardous	N
Usage not hard	N
Usage hard	N
Usage extra hard	Y
Usage attached to appliance	N
Usage pendant or portable	N
Usage refrigerators room AC as permitted	N
Usage electric vehicle charging	Y
Usage elevator	N
Insulation type	Thermoplastic with optional nylon
Insulation thickness range	30/Nylon 20 mils
Outer covering	Thermoset
Outer covering flame retardant	N
Outer covering moisture resistant	N
Outer braid on each conductor	N

Table 19-62 Flexible Cords Type EVJE

Flexible Cord and Cable	Feature Included
Type letter	EVJE
Voltage rating	300
AWG range	18-12
Number of conductors	2 or more + grounding
Location dry	N
Location wet	Y
Location hazardous	N
Usage not hard	Y
Usage hard	N

Table 19-62 *(Continued)*

Flexible Cord and Cable	Feature Included
Usage extra hard	N
Usage attached to appliance	N
Usage pendant or portable	N
Usage refrigerators room AC as permitted	N
Usage electric vehicle charging	Y
Usage elevator	N
Insulation type	Thermoplastic elastomer with optional nylon
Insulation thickness range	30 Nylon 20 mils
Outer covering	Thermoplastic elastomer
Outer covering flame retardant	N
Outer covering moisture resistant	N
Outer braid on each conductor	Optional

Table 19-63 Flexible Cords Type EVT

Flexible Cord and Cable	Feature Included
Type letter	EVT
Voltage rating	600
AWG range	18-500
Number of conductors	2 or more plus grounding
Location dry	N
Location wet	Y
Location hazardous	N
Usage not hard	N
Usage hard	N
Usage extra hard	Y
Usage attached to appliance	N
Usage pendant or portable	N
Usage refrigerators room AC as permitted	N
Usage electric vehicle charging	Y
Usage elevator	N

(continued)

Table 19-63 *(Continued)*

Flexible Cord and Cable	Feature Included
Insulation type	Thermoplastic with optional nylon
Insulation thickness range	30-45-60-80-95 Nylon 20-30-45-60-75 mils
Outer covering	Thermoplastic
Outer covering flame retardant	N
Outer covering moisture resistant	N
Outer braid on each conductor	Optional

Table 19-64 Flexible Cords Type EVJT

Flexible Cord and Cable	Feature Included
Type letter	EVJT
Voltage rating	300
AWG range	18-12
Number of conductors	2 or more plus grounding
Location dry	N
Location wet	Y
Location hazardous	N
Usage not hard	N
Usage hard	Y
Usage extra hard	N
Usage attached to appliance	N
Usage pendant or portable	N
Usage refrigerators room AC as permitted	N
Usage electric vehicle charging	Y
Usage elevator	N
Insulation type	Thermoplastic with optional nylon
Insulation thickness range	30 Nylon 20 mils
Outer covering	Thermoplastic
Outer covering flame retardant	N
Outer covering moisture resistant	N
Outer braid on each conductor	Optional

Table 19-65 Flexible Cords Type G

Flexible Cord and Cable	Feature Included
Type letter	G
Voltage rating	2000
AWG range	12-500
Number of conductors	2 to 6 plus grounding conductor (s)
Location dry	N
Location wet	N
Location hazardous	N
Usage not hard	N
Usage hard	N
Usage extra hard	Y
Usage attached to appliance	N
Usage pendant or portable	Y
Usage refrigerators room AC as permitted	N
Usage electric vehicle charging	N
Usage elevator	N
Insulation type	Thermoset
Insulation thickness range	60-80-95 mils
Outer covering	Oil resistant thermoset
Outer covering flame retardant	N
Outer covering moisture resistant	N
Outer braid on each conductor	N

Table 19-66 Flexible Cord Type G-GC

Flexible Cord and Cable	Feature Included
Type letter	G-GC
Voltage rating	2000
AWG range	12-500
Number of conductors	3 to 6 plus grounding conductors plus ground check conductor
Location dry	N

(continued)

Table 19-66 (Continued)

Flexible Cord and Cable	Feature Included
Location wet	N
Location hazardous	N
Usage not hard	N
Usage hard	N
Usage extra hard	Y
Usage attached to appliance	N
Usage pendant or portable	Y
Usage refrigerators room AC as permitted	N
Usage electric vehicle charging	N
Usage elevator	N
Insulation type	Thermoset
Insulation thickness range	60-80-95 mils
Outer covering	Oil resistant thermoset
Outer covering flame retardant	N
Outer covering moisture resistant	N
Outer braid on each conductor	N

Table 19-67 Flexible Cord Type HFD

Flexible Cord and Cable	Feature Included
Type letter	HFD
Voltage rating	300
AWG range	18-12
Number of conductors	2 or 3
Location dry	Y
Location wet	N
Location hazardous	N
Usage not hard	Y
Usage hard	N
Usage extra hard	N
Usage attached to appliance	N

Table 19-67 (Continued)

Flexible Cord and Cable	Feature Included
Usage pendant or portable	Portable heaters
Usage refrigerators room AC as permitted	N
Usage electric vehicle charging	N
Usage elevator	N
Insulation type	Thermoset
Insulation thickness range	15 – 30 mils
Outer covering	Cotton or nylon
Outer covering flame retardant	N
Outer covering moisture resistant	N
Outer braid on each conductor	N

Table 19-68 Flexible Cord Type HPN

Flexible Cord and Cable	Feature Included
Type letter	HPN
Voltage rating	300
AWG range	18-12
Number of conductors	2 or 3
Location dry	N
Location wet	Damp
Location hazardous	N
Usage not hard	Y
Usage hard	N
Usage extra hard	N
Usage attached to appliance	N
Usage pendant or portable	Y
Usage refrigerators room AC as permitted	N
Usage electric vehicle charging	N
Usage elevator	N
Insulation type	Oil resistant thermoset

(continued)

Table 19-68 (Continued)

Flexible Cord and Cable	Feature Included
Insulation thickness range	45-60-95 mils
Outer covering	N
Outer covering flame retardant	N
Outer covering moisture resistant	N
Outer braid on each conductor	N

Table 19-69 Flexible Cord Type S

Flexible Cord and Cable	Feature Included
Type letter	S
Voltage rating	600
AWG range	18-2
Number of conductors	2 or more
Location dry	N
Location wet	Damp
Location hazardous	N
Usage not hard	N
Usage hard	N
Usage extra hard	Y
Usage attached to appliance	N
Usage pendant or portable	Y
Usage refrigerators room AC as permitted	N
Usage electric vehicle charging	N
Usage elevator	N
Insulation type	Thermoset
Insulation thickness range	30-45-60 mils
Outer covering	Thermoset
Outer covering flame retardant	N
Outer covering moisture resistant	N
Outer braid on each conductor	N

Table 19-70 Flexible Cord Type SC

Flexible Cord and Cable	Feature Included
Type letter	SC
Voltage rating	600
AWG range	8-250
Number of conductors	1 or more
Location dry	N
Location wet	N
Location hazardous	N
Usage not hard	N
Usage hard	N
Usage extra hard	Y
Usage attached to appliance	N
Usage pendant or portable	Portable
Usage refrigerators room AC as permitted	N
Usage electric vehicle charging	N
Usage elevator	N
Insulation type	Thermoset
Insulation thickness range	60-80-95 mils
Outer covering	N
Outer covering flame retardant	N
Outer covering moisture resistant	N
Outer braid on each conductor	N

Table 19-71 Flexible Cords Type SCE

Flexible Cord and Cable	Feature Included
Type letter	SCE
Voltage rating	600
AWG range	8-250
Number of conductors	One or more

(continued)

Table 19-71 *(Continued)*

Flexible Cord and Cable	Feature Included
Location dry	N
Location wet	N
Location hazardous	N
Usage not hard	N
Usage hard	N
Usage extra hard	Y
Usage attached to appliance	N
Usage pendant or portable	Portable
Usage refrigerators room AC as permitted	N
Usage electric vehicle charging	N
Usage elevator	N
Insulation type	Thermoplastic elastomer
Insulation thickness range	60-80-95 mils
Outer covering	Thermoplastic
Outer covering flame retardant	N
Outer covering moisture resistant	N
Outer braid on each conductor	N

Table 19-72 Flexible Cords Type SCT

Flexible Cord and Cable	Feature Included
Type letter	SCT
Voltage rating	600
AWG range	8-250
Number of conductors	One or more
Location dry	N
Location wet	N
Location hazardous	N
Usage not hard	N

Table 19-72 (Continued)

Flexible Cord and Cable	Feature Included
Usage hard	N
Usage extra hard	Y
Usage attached to appliance	N
Usage pendant or portable	Portable
Usage refrigerators room AC as permitted	N
Usage electric vehicle charging	N
Usage eevator	N
Insulation type	Thermoplastic
Insulation thickness range	60-80-95 mils
Outer covering	Thermoplastic
Outer covering flame retardant	N
Outer covering moisture resistant	N
Outer braid on each conductor	N

Table 19-73 Flexible Cord Type SE

Flexible Cord and Cable	Feature Included
Type letter	SE
Voltage rating	600
AWG range	18-2
Number of conductors	2 or more
Location dry	N
Location wet	Damp
Location hazardous	N
Usage not hard	N
Usage hard	N
Usage extra hard	Y
Usage attached to appliance	N

(continued)

Table 19-73 *(Continued)*

Flexible Cord and Cable	Feature Included
Usage pendant or portable	Y
Usage refrigerators room AC as permitted	N
Usage electric vehicle charging	N
Usage elevator	N
Insulation type	Thermoplastic elastomer
Insulation thickness range	30-45-60 mils
Outer covering	Thermoplastic elastomer
Outer covering flame retardant	N
Outer covering moisture resistant	N
Outer braid on each conductor	N

Table 19-74 Flexible Cord Type SEW

Flexible Cord and Cable	Feature Included
Type letter	SEW
Voltage rating	600
AWG range	18-2
Number of conductors	2 or more
Location dry	N
Location wet	Y and Damp
Location hazardous	N
Usage not hard	N
Usage hard	N
Usage extra hard	Y
Usage attached to appliance	N
Usage pendant or portable	Y
Usage refrigerators room AC as permitted	N
Usage electric vehicle charging	N
Usage elevator	N
Insulation type	Thermoplastic elastomer
Insulation thickness range	30-45-60 mils

Table 19-74 *(Continued)*

Flexible Cord and Cable	Feature Included
Outer covering	Thermoplastic elastomer
Outer covering flame retardant	N
Outer covering moisture resistant	N
Outer braid on each conductor	N

Table 19-75 Flexible Cords Type SEO

Flexible Cord and Cable	Feature Included
Type letter	SEO
Voltage rating	600
AWG range	18-2
Number of conductors	2 or more
Location dry	N
Location wet	Damp and wet
Location hazardous	N
Usage not hard	N
Usage hard	N
Usage extra hard	Y
Usage attached to appliance	N
Usage pendant or portable	Y
Usage refrigerators room AC as permitted	N
Usage electric vehicle charging	N
Usage elevator	N
Insulation type	Thermoplastic elastomer
Insulation thickness range	30-45-60 mils
Outer covering	N
Outer covering flame retardant	N
Outer covering moisture resistant	N
Outer braid on each conductor	N

Table 19-76 Flexible Cords Type SEOW

Flexible Cord and Cable	Feature Included
Type letter	SEOW
Voltage rating	600
AWG range	18-2
Number of conductors	2 or more
Location dry	N
Location wet	damp
Location hazardous	N
Usage not hard	N
Usage hard	N
Usage extra hard	Y
Usage attached to appliance	N
Usage pendant or portable	Y
Usage refrigerators room AC as permitted	N
Usage electric vehicle charging	N
Usage elevator	N
Insulation type	Thermoplastic elastomer
Insulation thickness range	30-45-60 mils
Outer covering	N
Outer covering flame retardant	N
Outer covering moisture resistant	N
Outer braid on each conductor	N

Table 19-77 Flexible Cord Type SEOO

Flexible Cord and Cable	Feature Included
Type letter	SEOO
Voltage rating	600
AWG range	NA
Number of conductors	NA

Table 19-77 *(Continued)*

Flexible Cord and Cable	Feature Included
Location dry	N
Location wet	Damp
Location hazardous	N
Usage not hard	NA
Usage hard	NA
Usage extra hard	NA
Usage attached to appliance	NA Theater stage and garage
Usage pendant or portable	NA
Usage refrigerators room AC as permitted	NA
Usage electric vehicle charging	N
Usage elevator	N
Insulation type	Oil resistant thermoplastic elastomer
Insulation thickness range in mils	NA
Outer covering	N
Outer covering flame retardant	N
Outer covering moisture resistant	N
Outer braid on each conductor	N

Table 19-78 Flexible Cord Type SEOOW

Flexible Cord and Cable	Feature Included
Type letter	SEOOW
Voltage rating	600
AWG range	NA
Number of conductors	NA
Location dry	N
Location wet	Damp and wet
Location hazardous	N
Usage not hard	N
Usage hard	N

(continued)

Table 19-78 *(Continued)*

Flexible Cord and Cable	Feature Included
Usage extra hard	N
Usage attached to appliance	N
Usage pendant or portable	N
Usage refrigerators room AC as permitted	N
Usage electric vehicle charging	N
Usage elevator	N
Insulation type	NA
Insulation thickness range in mils	NA
Outer covering	N
Outer covering flame retardant	N
Outer covering moisture resistant	N
Outer braid on each conductor	N

Table 19-79 Flexible Cord Type SJ

Flexible Cord and Cable	Feature Included
Type letter	SJ
Voltage rating	300
AWG range	18-10
Number of conductors	2 to 6
Location dry	N
Location wet	Damp
Location hazardous	N
Usage not hard	N
Usage hard	Y
Usage extra hard	N
Usage attached to appliance	N
Usage pendant or portable	Y
Usage refrigerators room AC as permitted	N
Usage electric vehicle charging	N

Table 19-79 *(Continued)*

Flexible Cord and Cable	Feature Included
Usage elevator	N
Insulation type	Thermoset
Insulation thickness range	30, # 10 AWG 45 mils
Outer covering	Thermoset
Outer covering flame retardant	N
Outer covering moisture resistant	N
Outer braid on each conductor	N

Table 19-80 Flexible Cord Type SJE

Flexible Cord and Cable	Feature Included
Type letter	SJE
Voltage rating	300
AWG range	18-10
Number of conductors	2 to 6
Location dry	N
Location wet	Damp
Location hazardous	N
Usage not hard	N
Usage hard	Y
Usage extra hard	N
Usage attached to appliance	N
Usage pendant or portable	Y
Usage refrigerators room AC as permitted	N
Usage electric vehicle charging	N
Usage elevator	N
Insulation type	Thermoplastic elastomer
Insulation thickness range	30, # 10 AWG 45 mils
Outer covering	Thermoplastic elastomer
Outer covering flame retardant	N
Outer covering moisture resistant	N
Outer braid on each conductor	N

Table 19-81 Flexible Cord Type SJEW

Flexible Cord and Cable	Feature Included
Type letter	SJEW
Voltage rating	300
AWG range	18-10
Number of conductors	2 to 6
Location dry	N
Location wet	Damp and wet
Location hazardous	N
Usage not hard	N
Usage hard	Y
Usage extra hard	N
Usage attached to appliance	N
Usage pendant or portable	Y
Usage refrigerators room AC as permitted	N
Usage electric vehicle charging	N
Usage elevator	N
Insulation type	Thermoplastic elastomer
Insulation thickness range	30, # 120 AWG 45 mils
Outer covering	Thermoplastic elastomer
Outer covering flame retardant	N
Outer covering moisture resistant	N
Outer braid on each conductor	N

Table 19-82 Flexible Cord Type SJEO

Flexible Cord and Cable	Feature Included
Type letter	SJEO
Voltage rating	300
AWG range	18-10
Number of conductors	2 to 6

Table 19-82 *(Continued)*

Flexible Cord and Cable	Feature Included
Location dry	N
Location wet	Damp
Location hazardous	N
Usage not hard	N
Usage hard	Y
Usage extra hard	N
Usage attached to appliance	N
Usage pendant or portable	N
Usage refrigerators room AC as permitted	N
Usage electric vehicle charging	N
Usage elevator	N
Insulation type	Thermoplastic elastomer
Insulation thickness range	30 # 10 AWG 45 mils
Outer covering	Oil resistant thermoplastic elastomer
Outer covering flame retardant	N
Outer covering moisture resistant	N
Outer braid on each conductor	N

Table 19-83 Flexible Cord Type SJEOW

Flexible Cord and Cable	Feature Included
Type letter	SJEOW
Voltage rating	300
AWG range	18-10
Number of conductors	2 to 6
Location dry	N
Location wet	Damp and wet
Location hazardous	N
Usage not hard	N
Usage hard	Y

(continued)

Table 19-83 *(Continued)*

Flexible Cord and Cable	Feature Included
Usage extra hard	N
Usage attached to appliance	N
Usage pendant or portable	Y
Usage refrigerators room AC as permitted	N
Usage electric vehicle charging	N
Usage elevator	N
Insulation type	Thermoplastic elastomer
Insulation thickness range	30 # 10 AWG 45 mils
Outer covering	Oil resistant thermoplastic elastomer
Outer covering flame retardant	N
Outer covering moisture resistant	N
Outer braid on each conductor	N

Table 19-84 Flexible Cord Type SJO

Flexible Cord and Cable	Feature Included
Type letter	SJO
Voltage rating	300
AWG range	18-10
Number of conductors	2 to 6
Location dry	N
Location wet	Damp
Location hazardous	N
Usage not hard	N
Usage hard	Y
Usage extra hard	N
Usage attached to appliance	N
Usage pendant or portable	Y
Usage refrigerators room AC as permitted	N
Usage electric vehicle charging	N
Usage elevator	N
Insulation type	Thermoset

Table 19-84 *(Continued)*

Flexible Cord and Cable	Feature Included
Insulation thickness range	30, # 10 AWG 45 mils
Outer covering	Oil resistant thermoplastic elastomer
Outer covering flame retardant	N
Outer covering moisture resistant	N
Outer braid on each conductor	N

Table 19-85 Flexible Cord Type SJOW

Flexible Cord and Cable	Feature Included
Type letter	SJOW
Voltage rating	300
AWG range	18-10
Number of conductors	2 to 6
Location dry	N
Location wet	Damp and wet
Location hazardous	N
Usage not hard	N
Usage hard	Y
Usage extra hard	N
Usage attached to appliance	N
Usage pendant or portable	Y
Usage refrigerators room AC as permitted	N
Usage electric vehicle charging	N
Usage elevator	N
Insulation type	Thermoset
Insulation thickness range	30, # 10 AWG 45 mils
Outer covering	Oil resistant thermoset
Outer covering flame retardant	N
Outer covering moisture resistant	N
Outer braid on each conductor	N

Table 19-86 Flexible Cord Type SJT

Flexible Cord and Cable	Feature Included
Type letter	SJT
Voltage rating	300
AWG range	18-10
Number of conductors	2 to 6
Location dry	N
Location wet	Damp
Location hazardous	N
Usage not hard	N
Usage hard	Y
Usage extra hard	N
Usage attached to appliance	N
Usage pendant or portable	Y
Usage refrigerators room AC as permitted	N
Usage electric vehicle charging	N
Usage elevator	N
Insulation type	Thermoplastic
Insulation thickness range	30 # 10 AWG 45 mils
Outer covering	N
Outer covering flame retardant	N
Outer covering moisture resistant	N
Outer braid on each conductor	N

Table 19-87 Flexible Cord Type SJTO

Flexible Cord and Cable	Feature Included
Type letter	SJTO
Voltage rating	300
AWG range	18-10
Number of conductors	2 to 6
Location dry	N

Table 19-87 *(Continued)*

Flexible Cord and Cable	Feature Included
Location wet	Damp
Location hazardous	N
Usage not hard	N
Usage hard	Y
Usage extra hard	N
Usage attached to appliance	N
Usage pendant or portable	Y
Usage refrigerators room AC as permitted	N
Usage electric vehicle charging	N
Usage elevator	N
Insulation type	Thermoplastic
Insulation thickness range	30, # 10 AWG 45 mils
Outer covering	Oil resistant thermoplastic
Outer covering flame retardant	N
Outer covering moisture resistant	N
Outer braid on each conductor	N

Table 19-88 Flexible Cord Type SJTOO

Flexible Cord and Cable	Feature Included
Type letter	SJTOO
Voltage rating	300
AWG range	18-10
Number of conductors	2 to 6
Location dry	N
Location wet	Damp
Location hazardous	N
Usage not hard	N
Usage hard	Y
Usage extra hard	N
Usage attached to appliance	N

(continued)

Table 19-88 *(Continued)*

Flexible Cord and Cable	Feature Included
Usage pendant or portable	Y
Usage refrigerators room AC as permitted	N
Usage electric vehicle charging	N
Usage elevator	N
Insulation type	Oil resistant thermoplastic
Insulation thickness range	30 # 10 AWG 45 mils
Outer covering	Oil resistant thermoplastic
Outer covering flame retardant	N
Outer covering moisture resistant	N
Outer braid on each conductor	N

Table 19-89 Flexible Cord Type SJTOOW

Flexible Cord and Cable	Feature Included
Type letter	SJTOOW
Voltage rating	300
AWG range	18-10
Number of conductors	2 to 6
Location dry	N
Location wet	Damp and wet
Location hazardous	N
Usage not hard	N
Usage hard	Y
Usage extra hard	N
Usage attached to appliance	N
Usage pendant or portable	Y
Usage refrigerators room AC as permitted	N
Usage electric vehicle charging	N
Usage elevator	N
Insulation type	Oil resistant thermoplastic
Insulation thickness range	30, # 10 AWG 45 mils

Table 19-89 (Continued)

Flexible Cord and Cable	Feature Included
Outer covering	Oil resistant thermoplastic
Outer covering flame retardant	N
Outer covering moisture resistant	N
Outer braid on each conductor	N

Table 19-90 Flexible Cord Type SO

Flexible Cord and Cable	Feature Included
Type letter	SO
Voltage rating	600
AWG range	18-2
Number of conductors	2 or more
Location dry	N
Location wet	Damp
Location hazardous	N
Usage not hard	N
Usage hard	N
Usage extra hard	Y
Usage attached to appliance	N
Usage pendant or portable	Y
Usage refrigerators room AC as permitted	N
Usage electric vehicle charging	N
Usage elevator	N
Insulation type	Thermoset
Insulation thickness range	30 mils
Outer covering	Oil resistant thermoset
Outer covering flame retardant	N
Outer covering moisture resistant	N
Outer braid on each conductor	N

Table 19-91 Flexible Cord Type SOW

Flexible Cord and Cable	Feature Included
Type letter	SOW
Voltage rating	600
AWG range	18-2
Number of conductors	2 or more
Location dry	N
Location wet	Damp and wet
Location hazardous	N
Usage not hard	N
Usage hard	N
Usage extra hard	Y
Usage attached to appliance	N
Usage pendant or portable	Y
Usage refrigerators room AC as permitted	N
Usage electric vehicle charging	N
Usage elevator	N
Insulation type	Thermoset
Insulation thickness range	30 mils
Outer covering	Oil resistant thermoset
Outer covering flame retardant	N
Outer covering moisture resistant	N
Outer braid on each conductor	N

Table 19-92 Flexible Cord Type SPE-3

Flexible Cord and Cable	Feature Included
Type letter	SPE-3
Voltage rating	300
AWG range	18-10
Number of conductors	2 or 3
Location dry	N

Table 19-92 (Continued)

Flexible Cord and Cable	Feature Included
Location wet	Damp
Location hazardous	N
Usage not hard	Y
Usage hard	N
Usage extra hard	N
Usage attached to appliance	N
Usage pendant or portable	N
Usage refrigerators room AC as permitted	Y
Usage electric vehicle charging	N
Usage elevator	N
Insulation type	Thermoplastic elastomer
Insulation thickness range	60, 80, 95, 110 mils
Outer covering	N
Outer covering flame retardant	N
Outer covering moisture resistant	N
Outer braid on each conductor	N

Table 19-93 Flexible Cord Type SPT-3

Flexible Cord and Cable	Feature Included
Type letter	SPT-3
Voltage rating	300
AWG range	18-10
Number of conductors	2 or 3
Location dry	N
Location wet	Damp
Location hazardous	N
Usage not hard	Y
Usage hard	N
Usage extra hard	N
Usage attached to appliance	N

(continued)

Table 19-93 *(Continued)*

Flexible Cord and Cable	Feature Included
Usage pendant or portable	N
Usage refrigerators room AC as permitted	Y
Usage electric vehicle charging	N
Usage elevator	N
Insulation type	Thermoplastic
Insulation thickness range	60, 80, 95, 110 mils
Outer covering	N
Outer covering flame retardant	N
Outer covering moisture resistant	N
Outer braid on each conductor	N

Table 19-94 Flexible Cord Type SRD

Flexible Cord and Cable	Feature Included
Type letter	SRD
Voltage rating	300
AWG range	10-4
Number of conductors	3 or 4
Location dry	N
Location wet	Damp
Location hazardous	N
Usage not hard	N
Usage hard	N
Usage extra hard	N
Usage attached to appliance	Ranges, dryers
Usage pendant or portable	Portable
Usage refrigerators room AC as permitted	N
Usage electric vehicle charging	N
Usage elevator	N
Insulation type	Thermoset
Insulation thickness range	45 mils

Table 19-94 *(Continued)*

Flexible Cord and Cable	Feature Included
Outer covering	Thermoset
Outer covering flame retardant	N
Outer covering moisture resistant	N
Outer braid on each conductor	N

Table 19-95 Flexible Cord Type St

Flexible Cord and Cable	Feature Included
Type letter	ST
Voltage rating	600
AWG range	18-2
Number of conductors	2 or more
Location dry	N
Location wet	Damp
Location hazardous	N
Usage not hard	N
Usage hard	N
Usage extra hard	Y
Usage attached to appliance	Ranges, dryers
Usage pendant or portable	Y
Usage refrigerators room AC as permitted	N
Usage electric vehicle charging	N
Usage elevator	N
Insulation type	Thermoplastic
Insulation thickness range	30, 45, 60 mils
Outer covering	N
Outer covering flame retardant	N
Outer covering moisture resistant	N
Outer braid on each conductor	N

Table 19-96 Flexible Cord Type STW

Flexible Cord and Cable	Feature Included
Type letter	STW
Voltage rating	600
AWG range	18-2
Number of conductors	2 or more
Location dry	N
Location wet	Damp and wet
Location hazardous	N
Usage not hard	N
Usage hard	N
Usage extra hard	Y
Usage attached to appliance	N
Usage pendant or portable	Y
Usage refrigerators room AC as permitted	N
Usage electric vehicle charging	N
Usage elevator	N
Insulation type	Thermoplastic
Insulation thickness range	30, 45, 60 mils
Outer covering	N
Outer covering flame retardant	N
Outer covering moisture resistant	N
Outer braid on each conductor	N

Table 19-97 Flexible Cord Type STO

Flexible Cord and Cable	Feature Included
Type letter	STO
Voltage rating	600
AWG range	18-2
Number of conductors	2 or more
Location dry	N

Table 19-97 *(Continued)*

Flexible Cord and Cable	Feature Included
Location wet	Damp
Location hazardous	N
Usage not hard	N
Usage hard	N
Usage extra hard	Y
Usage attached to appliance	N
Usage pendant or portable	Y
Usage refrigerators room AC as permitted	N
Usage electric vehicle charging	N
Usage elevator	N
Insulation type	Thermoplastic
Insulation thickness range	30, 45, 60 mils
Outer covering	Oil resistant thermoplastic
Outer covering flame retardant	N
Outer covering moisture resistant	N
Outer braid on each conductor	N

Table 19-98 Flexible Cord Type STOOW

Flexible Cord and Cable	Feature Included
Type letter	STOOW
Voltage rating	600
AWG range	18-2
Number of conductors	2 or more
Location dry	N
Location wet	Damp
Location hazardous	N
Usage nNot hard	N
Usage hard	N
Usage extra hard	Y
Usage attached to appliance	N

Table 19-98 (Continued)

Flexible Cord and Cable	Feature Included
Usage pendant or portable	Y
Usage refrigerators room AC as permitted	N
Usage electric vehicle charging	N
Usage elevator	N
Insulation type	Oil resistant thermoplastic
Insulation thickness range	30, 45, 60 mils
Outer covering	N
Outer covering flame retardant	N
Outer covering moisture resistant	N
Outer braid on each conductor	N

Table 19-99 Flexible Cord Type SV

Flexible Cord and Cable	Feature Included
Type letter	SV
Voltage rating	300
AWG range	18-16
Number of conductors	2 or 3
Location dry	N
Location wet	Damp
Location hazardous	N
Usage not hard	Y
Usage hard	N
Usage extra hard	N
Usage attached to appliance	N
Usage pendant or portable	Y
Usage refrigerators room AC as permitted	N
Usage electric vehicle charging	N
Usage elevator	N
Insulation type	Thermoset

Table 19-99 *(Continued)*

Flexible Cord and Cable	Feature Included
Insulation thickness range	15 mils
Outer covering	Thermoset
Outer covering flame retardant	N
Outer covering moisture resistant	N
Outer braid on each conductor	N

Table 19-100 Flexible Cord Type SVE

Flexible Cord and Cable	Feature Included
Type letter	SVE
Voltage rating	300
AWG range	18-16
Number of conductors	2 or 3
Location dry	N
Location wet	Damp
Location hazardous	N
Usage not hard	Y
Usage hard	N
Usage extra hard	N
Usage attached to appliance	N
Usage pendant or portable	Y
Usage refrigerators room AC as permitted	N
Usage electric vehicle charging	N
Usage elevator	N
Insulation type	Thermoplastic elastomer
Insulation thickness range	15 mils
Outer covering	Thermoplastic
Outer covering flame retardant	N
Outer covering moisture resistant	N
Outer braid on each conductor	N

Table 19-101 Flexible Cord Type SVEO

Flexible Cord and Cable	Feature Included
Type letter	SVEO
Voltage rating	300
AWG range	18-16
Number of conductors	2 or 3
Location dry	N
Location wet	Damp
Location hazardous	N
Usage not hard	Y
Usage hard	N
Usage extra hard	N
Usage attached to appliance	N
Usage pendant or portable	Y
Usage refrigerators room AC as permitted	N
Usage electric vehicle charging	N
Usage elevator	N
Insulation type	Thermoplastic elastomer
Insulation thickness range	15 mils
Outer covering	Oil resistant thermoplastic
Outer covering flame retardant	N
Outer covering moisture resistant	N
Outer braid on each conductor	N

Table 19-102 Flexible Cord Type SVEOO

Flexible Cord and Cable	Feature Included
Type letter	SVEOO
Voltage rating	300
AWG range	18-16
Number of conductors	2 or 3
Location dry	N

Table 19-102 (Continued)

Flexible Cord and Cable	Feature Included
Location wet	Damp
Location hazardous	N
Usage not hard	Y
Usage hard	N
Usage extra hard	N
Usage attached to appliance	N
Usage pendant or portable	Y
Usage refrigerators room AC as permitted	N
Usage electric vehicle charging	N
Usage elevator	N
Insulation type	Oil resistant thermoplastic elastomer
Insulation thickness range	15 mils
Outer covering	Oil resistant thermoplastic elastomer
Outer covering flame retardant	N
Outer covering moisture resistant	N
Outer braid on each conductor	N

Table 19-103 Flexible Cord Type TPT

Flexible Cord and Cable	Feature Included
Type letter	TPT
Voltage rating	300
AWG range	27
Number of conductors	2
Location dry	N
Location wet	Damp
Location hazardous	N
Usage not hard	Y
Usage hard	N

(continued)

Table 19-103 *(Continued)*

Usage extra hard	N
Usage attached to appliance	Y
Usage pendant or portable	N
Usage refrigerators room AC as permitted	N
Usage electric vehicle charging	N
Usage elevator	N
Insulation type	Thermoplastic
Insulation thickness range	30 mils
Outer covering	Thermoplastic
Outer covering flame retardant	N
Outer covering moisture resistant	N
Outer braid on each conductor	N

Table 19-104 Flexible Cord Type TST

Flexible Cord and Cable	Feature Included
Type letter	TST
Voltage rating	300
AWG range	27
Number of conductors	2
Location dry	N
Location wet	Damp
Location hazardous	N
Usage not hard	Y
Usage hard	N
Usage extra hard	N
Usage attached to appliance	Y
Usage pendant or portable	N
Usage refrigerators room AC as permitted	N
Usage electric vehicle charging	N
Usage elevator	N
Insulation type	Thermoset

Table 19-104 (Continued)

Flexible Cord and Cable	Feature Included
Insulation thickness range	15 mils
Outer covering	Thermoplastic
Outer covering flame retardant	Thermoplastic
Outer covering moisture resistant	N
Outer braid on each conductor	N

Table 19-105 Flexible Cord Type W

Flexible Cord and Cable	Feature Included
Type letter	W
Voltage rating	2000
AWG range	12-500. 5-1 to 1000 KCM
Number of conductors	1 to 6
Location dry	N
Location wet	N
Location hazardous	N
Usage not hard	N
Usage hard	N
Usage extra hard	Y
Usage attached to appliance	N
Usage pendant or portable	Portable
Usage refrigerators room AC as permitted	N
Usage electric vehicle charging	N
Usage elevator	N
Insulation type	Thermoset
Insulation thickness range	60, 80, 95, 110 mils
Outer covering	Oil resistant thermoset
Outer covering flame retardant	N
Outer covering moisture resistant	N
Outer braid on each conductor	N

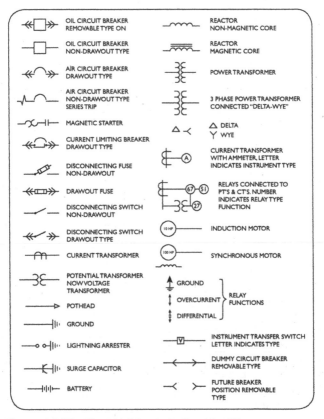

OIL CIRCUIT BREAKER REMOVABLE TYPE ON	REACTOR NON-MAGNETIC CORE
OIL CIRCUIT BREAKER NON-DRAWOUT TYPE	REACTOR MAGNETIC CORE
AIR CIRCUIT BREAKER DRAWOUT TYPE	POWER TRANSFORMER
AIR CIRCUIT BREAKER NON-DRAWOUT TYPE SERIES TRIP	3 PHASE POWER TRANSFORMER CONNECTED "DELTA-WYE"
MAGNETIC STARTER	△ DELTA Y WYE
CURRENT LIMITING BREAKER DRAWOUT TYPE	CURRENT TRANSFORMER WITH AMMETER, LETTER INDICATES INSTRUMENT TYPE
DISCONNECTING FUSE NON-DRAWOUT	RELAYS CONNECTED TO PT'S & CT'S. NUMBER INDICATES RELAY TYPE FUNCTION
DRAWOUT FUSE	
DISCONNECTING SWITCH NON-DRAWOUT	
DISCONNECTING SWITCH DRAWOUT TYPE	INDUCTION MOTOR
CURRENT TRANSFORMER	SYNCHRONOUS MOTOR
POTENTIAL TRANSFORMER NOW VOLTAGE TRANSFORMER	GROUND OVERCURRENT RELAY FUNCTIONS DIFFERENTIAL
POTHEAD	
GROUND	INSTRUMENT TRANSFER SWITCH LETTER INDICATES TYPE
LIGHTNING ARRESTER	DUMMY CIRCUIT BREAKER REMOVABLE TYPE
SURGE CAPACITOR	FUTURE BREAKER POSITION REMOVABLE TYPE
BATTERY	

Figure 19-50 Electrical Symbols One-Line Diagrams

Table 19-106 Cable and Conductor Insulation Type and Application Letter Designations

Cable and Conductor Insulation Type and Application Letter Designations	Features
A	Asbestos. No new installations, once used for high temperature applications
AWM	Appliance wiring material, listed for 105 °C. Typically PVC or thermoplastic with nylon jacket. THHN may be listed as AWM.
C	Cotton
CIC	Factory-made cable in conduit, also known as duct cable
CPE	Chlorinated polyethylene Jacket
E	Ethylene
E	Elastomer
E	Thermoset elevator cable with flexible nylon jacket cotton braids and outer covering
EO	Thermoset elevator cable with cotton braids and outer covering
EPR	Ethylene propylene rubber
ETT	Thermoplastic elevator cable, no outer covering, listed for hazardous locations
ETP	Thermoplastic elevator cable with rayon braid, hazardous locations
ETFE	Ethylene tertrofluoroethylene
EV	
EVE	Thermoset electric vehicle cable, wet locations, extra hard usage, 600 V
EVJ	Thermoplastic elastomer eclectic vehicle cable, wet locations, extra hard usage, 600 V
EVJT	Thermoset electric vehicle cable, wet locations, hard usage, 300 V
	Thermoplastic electric vehicle cable, wet locations, hard usage, 300 V

(continued)

Table 19-106 *(Continued)*

Cable and Conductor Insulation Type and Application Letter Designations	Features
F	Fluorinated, FFH-2
FEP	Fluorinated ethylene propylene
G G	Glass braids are lacquered or saturated with moisture-resistant material. May be fibered glass yarn or cotton, rayon or silk. Thermoset oil-resistant, portable power cable, listed to 2000 V, portable and extra hard usage
H HPD	Without the letter "H" the material is listed for a maximum temperature of 60 °C. With one H 75 °C, with two H, listed for 90 °C Thermoset heater cord, 300 V, cotton or rayon outer covering, dry location, not hard usage
HPN	Thermoset oil-resistant parallel heater cord, damp locations, 300 V
HSJ	Thermoset jacketed heater cord, 300 V damp locations, hard usage, cotton and thermoset outer covering, portable heater
HSJO	
HSJOO	Thermoset jacketed heater cord, cotton and oil-resistant thermoset outer covering, 300 V damp locations, hard usage
HFF	Oil-resistant thermoset insulation, cotton oil-resistant outer covering, damp locations, portable heater, 300 V, hard usage Halar Ethylene Chrorotrifluorothylene, fixture wire
J	See SJ
KF-1	Kapton fixture wire, aromatic polyimide. Listed to 200 °C. Listed to 300 V

Table 19-106 *(Continued)*

Cable and Conductor Insulation Type and Application Letter Designations	Features
MC MCHL MI MGR, MGT MWT	Metal clad cable Metal clad cable listed for classified (Hazardous) locations Magnesium oxide with copper tubing outer jacket, listed for use in classified (Hazardous) locations Special oven wire may be oil resistant, not for high flexing applications Machine tool wire
N NISP-1& 2 NISPE-1 NISPT-1 & 2 NM-B NM-C NPLC	Nylon, or equal Nonintegral parallel cords, thermoset 300 V, damp locations, pendant or portable, not hard usage Nonintegral parallel cords, thermoplastic elastomer insulation, outer covering of thermoplastic elastomer, 300 V, pendant or portable, damp locations, not hard usage Nonintegral parallel cords, thermoplastic insulation, with thermoplastic outer covering, 300 V, pendant or portable usage, damp locations, hot hard usage Nonmetallic sheathed cable, B indicated individual conductors listed for 90 °C with ampacity limited to 60 °C. Thermoplastic, (typically PVC) with outer nylon jacket Nonmetallic sheathed cable Non-power-limited fire protection cable
O OO	Oil resistant and oil resistant II = 75 °C, I = 60 °C, Gasoline and oil resistant II = 75 °C, I = 60 °C Conductors and jacket oil resistant

(continued)

Table 19-106 *(Continued)*

Cable and Conductor Insulation Type and Application Letter Designations	Features
P PD PFA PLTC PPE PPP or PPLPP PTF and PTFF PF & PFG PFF & PGFF	Can represent Propylene or Paper, or Perflouroalkoxy Twisted portable power cords, thermoplastic or thermoset, insulation, cotton or rayon outer covering listed to 2000 V, pendant or portable usage, dry locations, not hard usage Perflouroalkoxy Power-limited tray cable Portable power cable with thermoplastic elastomer insulation, with outer covering oil-resistant thermoplastic elastomer, portable or extra hard usage Paper with polypropylene transmission cable Fixture wire listed for 250 °C and 150 °C, respectively, Polyimide tape insulation, see also KF-1 Listed to 200 °C, 600 V Listed to 150 °C, 600 V
R RFFH-1, RFHH-3, RFH-2, FFH-2 RUH	Thermoset material rubber or equal (synthetic neoprene) RFHH-1 and RFHH-3 to 90 °C, RHF-2 and FFH-2 to 90 °C, all 600 V 90% latex, direct burial and dry locations to 140 °F

Table 19-106 *(Continued)*

Cable and Conductor Insulation Type and Application Letter Designations	Features
S	Silicone
S	Hard service, flexible cable, thermoset insulation, with thermoset outer covering, pendant or portable usage, damp locations, extra hard usage, 600 V
SC	Flexible stage and lighting power cord, thermoset outer covering, 600 V, portable, extra hard usage
SE	Hard service, thermoplastic elastomer insulation, pendant or portable, damp locations, 600 V, usage at various temperatures. Sunlight resistant but not marked on cable. Extra hard usage
SF-1 & SF-2	Listed to 200 °C, 300 and 600 V, respectively
SFF-1 & SFF-2	Listed to 150 °C, 300 and 600 V, respectively
SIS	Silicone rubber
SJ	Junior hard service flexible cable, 300 V, thermoset insulation, with thermoset outer covering, pendant and portable usage, damp locations, hard usage
SP	Thermoset parallel cord, thermoset insulation, damp locations, 300 V, pendant or portable, not hard usage
SP-3	Refrigerators, room air conditioners, damp locations, not hard usage
SRD	Range dryer cable, thermoset insulation with thermoset outer covering, 300 V, portable, damp locations
SV	Vacuum cleaner cord, thermoset insulation 300 V, with thermoset outer covering, pendant or portable usage, damp locations, not hard usage

(continued)

Table 19-106 (Continued)

Cable and Conductor Insulation Type and Application Letter Designations	Features
T	Thermoplastic typically PVC. Thermoplastic will become more plastic-like when heated. Thermoset will retain its rigidity when repeatedly heated. Will char, but will not melt into a liquid-like material.
TA	
TC	Thermoplastic with asbestos fibers listed to 194 °F
TFE	Tray cable
TFN, TFFN	Extruded polyerafluoroethylene (Teflon) may be listed 400 to 900 °F applications
TF and TFF	PVC material with nylon (or equivalent) jacket. May be marked as oil and gasoline resistant. Nylon jacket minimum 4 mils thickness.
TPT	Flouroethylene, TF 60 °C, TFC 90 °C, TFF 60 °C, TFFN 90 °C, all 600 V
TPPO	Parallel tinsel cord with thermoplastic insulation, with thermoplastic outer covering, 300 V, damp location, not hard usage
	Thermoplastic polyolefin
UF	Underground feeder cable, listed for 60 °C, not sunlight resistant
USE	
UF-b	Underground service entrance cable, 75 °C wet location, moisture and heat resistant, with a -2 listed for 90 °C in wet locations
V	

Table 19-106 (Continued)

Cable and Conductor Insulation Type and Application Letter Designations	Features
W W W	Suitable for wet locations, but not continually submerged. Moisture resistant wet locations Portable power cord, with thermoset insulation, with oil-resistant thermoset outer covering, listed to 2000 V, portable, extra hard usage
X XF and XFP XLPE	Cross-linked synthetic polymer material Listed to 150 °C, 300 V. Materials same as XLPO Cross-linked polyethylene, fluorinated ethylene propylene (or polyolefin)
Z ZHF, ZF, and ZFF ZW	Modified ethylene tetrafluorethylene (ETFE) dry locations Listed to 200 °C, and 150 °C for ZF and ZFF, all 600 V Same as above, listed for wet locations
-2	Indicates listed for 90 °C in wet locations

Table 19-107 Fixture Wire DC Resistance per 1000 Ft, Stranded

Fixture Wire DC Resistance per 1000 Ft Uncoated Stranded Copper	At 68 °F	At 77 °F
18 AWG	6.66	6.79
16 AWG	4.18	4.26
14 AWG	2.62	2.67
12 AWG	1.65	1.68
10 AWG	1.04	1.06

Table 19-108 Wire Way Fill Table, Various Enclosures

Conduit, Tubing, Panel, Raceway or Wire Way Type	Total Conductor Cross-Sectional Area Not to Exceed 20% of Enclosure Cross-Sectional Area	Total Conductor Cross-Sectional Area Not to Exceed 40% of Enclosure Cross-Sectional Area	Splices and Taps Not to Exceed 75% of Enclosure Cross-Sectional Area	Maximum Fill for a Single Conductor in a Conduit or Tubing 53%	Maximum Fill for Two Conductors in a Conduit or Tubing 31%	Maximum Fill for Three or More Conductors in Conduit or Tubing 40%
Cellular metal floor raceways		X				
Metal wire ways	X					
Nonmetallic wire ways	X					
Industrial control panels		X	X			
Conduit and tubing				X	X	X

THREE PHASE MOTOR REFERENCE MATERIAL

Number of Poles and the Speed of the Motor

Two poles, 3600 RPMs synchronous, 3450 actual

Four poles, 1800 RPMs synchronous, 1725 actual

Six poles, 1200 RPMs synchronous, 1140 actual

Eight poles, 900 RPMs synchronous, 850 actual

Assumes 60 Hz

Stator Winding Electrical-Magnetic Wire Insulation Class

The copper wire used to make coils of wire that make-up the stator windings are coated with an enamel, or varnish like material that is designed to provide the needed electrical separation of the individual coils. In the U.S.A. this insulating material is divided into four classes. Each class of insulation has a maximum design operating temperature limit. The following table lists this temperature for each insulation class:

Class A, 105 C. Seldom used by the major motor manufactures today.

Class B, 130 C. for 1.15 SF 135 C

Class F, 150 C. for 1.15 SF 160 C

Class H, 180 C.

Class E, used by IEC motor manufactures, 120 C

The insulation classes are separated from each other by 25 degrees C, or 45 degrees F. When this insulation is operated at no more than and no less than this temperature it will have a design operating (service) life of some 20,000 hours and 5,000 starts. When the installation is operated at a temperature below the maximum, its life will increase. When the installation is operated at a temperature above this maximum, its life will be decreased.

This maximum temperature is very important. Heat kills electric motors. Operating a motor at just 10°C, or 18°F above its rated maximum temperature will result in a 50% decrease in the installations operating life. Just operating a motor 18 degrees F. hotter will decrease the life of a motor from twenty to ten years. That works out to a decrease of about 600 hours for every 1 1/8 degree F. above the maximum rated temperature.

Table 19-109 Three Phase Motor-NEC Locked Rotor Code

LRA Code Letter	KVA/HP Range	Midrange Value
A	0.00-3.14	1.6
B	3.15-3.54	3.3
C	3.55-3.99	3.8
D	4.00-4.49	4.3
F	5.00-5.59	4.7
G	5.60-6.29	5.3
H	6.30-7.09	5.9
J	7.10-7.99	6.7
K	8.00-8.99	7.5
L	9.00-9.99	8.5
M	10.00-11.19	9.5
N	11.20-12.49	10.6
O	12.50-13.99	11.8
R	14.00-15.99	13.2

Table 19-110 Three Phase Motor Starting Method

% of Full Voltage	Voltage at Motor	Line Current	Motor Torque
Full voltage (XL)	100%	100%	100%
Autotransformer			
80% tap	80%	64%	64%
65% tap	65%	42%	42%
50% tap	50%	25%	25%
Primary Resistor			
80% tap	80%	80%	64%
65% tap	65%	65%	42%
50% Tap	50%	50%	25%

Table 19-111 Motor Re-Lubrication Frequency Table

Horsepower Range	Re-Lubrication Interval	Hours of Service Each Year
1/8 to7.5	5 yrs.	5,000 hours
10 to 40	3 yrs.	5,000 hours
50 to 100	1 yr.	5,000 hours
Up to 7.5	2 yrs.	Continuous
10 to 40	1 yr.	Continuous
50 to 100	9 months	Continuous
All	1 yr.	Seasonal service idle for 6 months or more
1/8 to 40	6 months	Seasonal service idle for 6 months or more
50 to 150	3 months	Seasonal service idle for 6 months of more

Table 19-112 Motor Re-Lubrication Table Based upon RMP, Frame Size, and Hours of Operation

RPM	Horsepower Range	Frame Size Range	8 Hrs. Per Day Operation	24 Hrs. Per Day Operation
3600	½ to 7 1/2	184T	6 Months	3 Months
	10 to 40	213 T to 286 TS	4 Months	2 Months
	50 to 150	324TS to 405 TS	4 Months	2 Months
1800	½ to 7 1/5	143T to 213T	18 Months	9 Months
	10 to 40	215T to 324T	9 Months	4 Months
	50 to 150	326T to 444TS	9 Months	4 Months
1200	½ to 7 1/2	145 T to 254T	24 Months	12 Months
	10 to 40	256T to 364 T	12 Months	6 Months
	50 to 150	365T to 445 T	12 Months	6 Months
900	½ to 7 1/2	182T to 256T	24 Months	12 Months
	10 to 40	284T to 365T	12 Months	6 Months
	50 to 150	404T to 445T	12 Months	6 Months

Bearings

Table 19-113 provides a listing of bearings commonly used with various frame sizes of electrical motors by one manufacturer. Not all manufacturers use the same size and type bearings.

Table 19-113 Types of Bearings

Bearings Used in Open Housings	Aluminum	Steel Housing
Frame size	Drive end	Opposite Drive End Bearing
56-56H	203	203
143T-145T	205, 206	203
182T-184T	207	205
213T-215T	208, 209	206
254T-256T	309	208
284T/TS-286t/TS	310	209
324T/TS-326T/TS	311	309
364T/TS-365T/TS	313	311
404T/TS-405T/TS	315	313
444T/TS-4445T/TS	318, 315	315
TEFC Cast Iron Housings		
143T-145T	205	203, 205
182T-184T	306	205, 306
213T-215T	307, 308	207, 307
254T-256T	309	208, 209
284T/TS-286T/TS	310	210
324T/TS—326T/TS	312	211
364T/TS-365T/TS	313	213
404T/TS-405T/TS	315	215
444T/TS-445T/TS	318	315
447T/TS-449T/TS	320	315

Maximum Sleeve Bearing Wear

Like all things sleeve bearings will wear over time. Table 19-114 lists minimum and maximum measurements on shaft size and the maximum amount of bearing wear.

Table 19-114 Bearing Wear

Motor Shaft Size	Factory Clearance in Inches		Approximate Maximum Wear in Inches for Sleeve Bearings
	Minimum	Maximum	
¾ to 1 inch	0.0015	0.0025	0.0035–0.004
1 to ¼ inches	0.003	0.004	0.005–0.006
1 ¼ to 2 inches	0.0035	0.005	0.007–0.008
2 to 2 1/2 inches	0.004	0.006	0.008–0.009
2 ½ to 3 inches	0.006	0.007	0.009–0.0105
3 to 4 inches	.007	0.008	0.010–0.0115
4 to 5 inches	0.008	0.009	0.-11–0.0125
5 to 6 inches	0.008	0.010	0.012–0.014

Motor Frame Size and Shaft End Play

Table 19-115 provides measurements for the maximum amount of shaft end play that can be allowed for both sleeve and ball bearings by motor frame sizes according to one manufacturer.

Table 19-115 Shaft End Play

Motor Frame Size	Sleeve Bearings		Ball Bearings	
	Minimum	Maximum	Minimum	Maximum
18	0.040	0.065	0.020	0.050
20	0.050	0.075	0.020	0.055
21	90.055	0.085	0.020	0.055
22	0.060	0.08	0.020	0.060
25	0.070	0.100	0.025	0.060
28	0.075	0.110	0.025	0.065
32	0.080	0.130	0.030	0.065
36	0.090	0.140	0.030	0.075
40	0.090	0.140	0.035	0.080
44	0.110	0.160	0.040	0.085

(continued)

Table 19-115 *(Continued)*

Motor Frame Size	Sleeve Bearings		Ball Bearings	
50	0.130	0.180	0.040	0.095
58	0.140	0.190	0.050	0.105
68	0.160	0.225	0.055	0.110

Bearing Protection

When operating conditions such as vibration, temperature, and noise increase, the bearing life will have decreased. Where instrumentation is provided to monitor these conditions on a continuous basis, the condition of the bearing can be determined. Electronic instruments have advanced to the point to where that virtually any physical condition can be monitored. Table 19-116 can provide a guide as to normal, alarm, and shutdown conditions of lubricant temperature.

Table 19-116 Bearing Conditions

Condition	Lubricant	Type
	Standard	Synthetic
Normal	176 F or lower	230 F or lower
Alarm	194 F	248 F
Shutdown	212 F	266 F

Vibration Tolerances

Velocity guide lines are:

0.444 or greater should be considered very rough

0.222 should be considered rough

0.111 should be considered slightly rough

0.555 should be considered to be fair

0.0277 should be considered to be good

0.0139 should be considered to be very good

0.0069 should be considered to be smooth

0.0035 should be considered to be very smooth

Velocity readings below 0.0030 should be viewed with caution as they would be seldom found in commercial or industrial facilities. Readings that do not range or deviate over time should cause one to question the accuracy of the reading.

Equipment Vibration Criteria

The following data can be used as a guide to the amount of vibration that is considered normal and acceptable:

Pumps 0.13 in/s

Centrifugal compressors 0.13 in/s

Fans, centrifugal and axial 0.09 in/s

Table 19-117 Vibration Alarm Levels in Inches per Second

Type of Equipment	Alarm Value
Blower-Fans	
Direct drive	0.325 IPS
Belt drive	0.370 IPS
Compressors	
Reciprocating	0.450 IPS
Screw	0.390 IPS
Centrifugal	0.300 IPS
Cooling Tower Drives	
Log shaft	0.500 IPS
Close coupled	0.300 IPS
Belt driven	0.400 IPS
Motor-Gen sets	
Direct drive	0.300 IPS
Belt drive	0.400 IPS
Chillers	

(continued)

Table 19-117 (Continued)

Type of Equipment	Alarm Value
Centrifugal	0.250 IPS
Reciprocating	0.400 IPS
Centrifugal pumps	
Integral shaft	0.225 IPS
Direct drive	0.290 IPS
Vertical	0.300 IPS

Velocity Severity Chart

Table 19-118 can be used as an aid to the decision making process of when to take a machine off line to make the necessary repairs.

Table 19-118 Class II Electrical Motors up to 15KW

Velocity in Inches per Second		
1.0+	Shutdown	Destructive forces evident, Bearing failure, metal fatigue likely
1.0	Dangerous	Destructive forces beginning, corrective action within a few days required to minimize amount of damage
.6	Very rough major defect	Correct within two weeks, monitor daily
.4	Rough	Corrective action within one month
.3	Fair	Corrective action is economically viable
.2	Good	Improve to lower readings and improve service life
.1	Smooth	Not economical to repair

Vibration Criteria For Equipment

Table 19-119 can be used as a general guide where manufacturer recommendations are unknown.

Table 19-119 General Vibration Criteria

Equipment	Maximum Allowable Vibration Peak to Peak Displacement in Mill (0.001 in)
Pumps 1800 RPM	2 mills
Pumps 3600 RPM	1 mil
Centrifugal compressors	1 mil
Fans centrifugal and axial	
Under 600 RPM	4 mills
600 to 1,000 RPMs	3 mills
1000 to 2,000 RPMS	2 mills
Over 2,000 RPMs	1 mill

Rotating Equipment Displacement Values

For most rotating equipment used in industry with RPMs within the range of 800 to 3600 RPMs, the displacement values listed in Table 19-120 can serve as a guide.

Table 19-120 Displacement Values for Rotating Equipment

Slow Speed	Medium Speed	High Speed
Immediate corrective action required		
15 mills	8 mills	2.5 mills
Normal		
6 mills	3.75 mills	1.8 mills
Fair to Good Condition		
2.5 mills	1.7 mills	0.4 mills

Minimum Electrical Resistance When Tested With a Meg Ohm Meter

Some firms in the continuous process industries and some engineering and electrical testing firms have as previously stated established go-no-go megohm and test voltage values.

The values in Table 19-121 are the recommendations of one motor manufacturer. When not required elsewhere, these values can serve as default values.

Table 19-121 Default Go-No-Go Values

Minimum Go-No-Go Resistance Values	Motor Nameplate Voltage
100,000 ohms	208 volts and less
200,000 ohms	208-240
300,000 ohms	240-600
1 Meg ohm	600-1,000
2 Meg ohm	1,000-2,400
3 Meg ohm	2,400-5,000

The IEEE standard 43 recommends the following as minimum acceptable values for insulation resistance.

Class A insulation 1.5

Class B insulation 2.0

Class F insulation 2.0

Class H insulation 2.0

Modern terminology is *insulation system* as opposed to the old term "insulation class".

INDEX